CLIMATE CHANGE AND ITS CAUSES, EFFECTS AND PREDICTION

# EPA REGULATION OF GREENHOUSE GASES

## CONSIDERATIONS AND OPTIONS

# CLIMATE CHANGE AND ITS CAUSES, EFFECTS AND PREDICTION

Additional books in this series can be found on Nova's website under the Series tab.

Additional E-books in this series can be found on Nova's website under the E-books tab.

CLIMATE CHANGE AND ITS CAUSES, EFFECTS AND PREDICTION

# EPA REGULATION OF GREENHOUSE GASES

## CONSIDERATIONS AND OPTIONS

CIANNI MARINO
AND
NICO COSTA
EDITORS

Nova Science Publishers, Inc.
*New York*

Copyright © 2012 by Nova Science Publishers, Inc.

**All rights reserved.** No part of this book may be reproduced, stored in a retrieval system or transmitted in any form or by any means: electronic, electrostatic, magnetic, tape, mechanical photocopying, recording or otherwise without the written permission of the Publisher.

For permission to use material from this book please contact us:
Telephone 631-231-7269; Fax 631-231-8175
Web Site: http://www.novapublishers.com

## NOTICE TO THE READER

The Publisher has taken reasonable care in the preparation of this book, but makes no expressed or implied warranty of any kind and assumes no responsibility for any errors or omissions. No liability is assumed for incidental or consequential damages in connection with or arising out of information contained in this book. The Publisher shall not be liable for any special, consequential, or exemplary damages resulting, in whole or in part, from the readers' use of, or reliance upon, this material. Any parts of this book based on government reports are so indicated and copyright is claimed for those parts to the extent applicable to compilations of such works.

Independent verification should be sought for any data, advice or recommendations contained in this book. In addition, no responsibility is assumed by the publisher for any injury and/or damage to persons or property arising from any methods, products, instructions, ideas or otherwise contained in this publication.

This publication is designed to provide accurate and authoritative information with regard to the subject matter covered herein. It is sold with the clear understanding that the Publisher is not engaged in rendering legal or any other professional services. If legal or any other expert assistance is required, the services of a competent person should be sought. FROM A DECLARATION OF PARTICIPANTS JOINTLY ADOPTED BY A COMMITTEE OF THE AMERICAN BAR ASSOCIATION AND A COMMITTEE OF PUBLISHERS.

Additional color graphics may be available in the e-book version of this book.

## Library of Congress Cataloging-in-Publication Data

EPA regulation of greenhouse gases : considerations and options / editors, Cianni Marino and Nico Costa.
    p. cm.
  Reports by James E. McCarthy and Larry Parker of Congressional Research Service and Regina A. McCarthy of EPA.
  Includes bibliographical references and index.
  ISBN 978-1-61470-729-5 (hardcover)
  1. Greenhouse gases--Law and legislation--United States. 2. Greenhouse gas mitigation--Law and legislation--United States. 3. Greenhouse gas mitigation--Political aspects--United States. 4. Environmental policy--Political aspects--United States. 5. United States. Environmental Protection Agency. 6. United States. Clean Air Act. I. McCarthy, James E. II. Marino, Cianni. III. Costa, Nico. IV. Parker, Larry, 1954- V. McCarthy, Gina.
  KF3812.E63 2011
  344.7304'6342--dc23
            2011024891

*Published by Nova Science Publishers, Inc. † New York*

# CONTENTS

| | | |
|---|---|---|
| **Preface** | | **vii** |
| **Chapter 1** | EPA Regulation of Greenhouse Gases: Congressional Responses and Options *James E. McCarthy and Larry Parker* | **1** |
| **Chapter 2** | EPA's BACT Guidance for Greenhouse Gases from Stationary Sources *Larry Parker and James E. McCarthy* | **23** |
| **Chapter 3** | Cars, Trucks, and Climate: EPA Regulation of Greenhouse Gases from Mobile Sources *James E. McCarthy* | **49** |
| **Chapter 4** | Climate Change: Potential Regulation of Stationary Greenhouse Gas Sources under the Clean Air Act *Larry Parker and James E. McCarthy* | **73** |
| **Chapter 5** | Statement of Regina A. McCarthy, Assistant Administrator, Office of Air and Radiation, U.S. Environmental Protection Agency Before the Committee on Environment and Public works U.S. Senate   March 4, 2010 *Regina A. McCarthy* | **113** |

| | | |
|---|---|---|
| **Chapter 6** | Statement of Regina A. McCarthy, Assistant Administrator, Office of Air and Radiation, U.S. Environmental Protection Agency Before the Subcommittee on Clean air and Nuclear safety Committee on Environment and Public works U.S. Senate July 9, 2009 *Regina A. McCarthy* | **119** |
| **Index** | | **127** |

# PREFACE

As a direct result of the Environmental Protection Agency's promulgation of an "endangerment finding" for greenhouse gas (GHG) emissions in December 2009, and its subsequent promulgation of GHG emission standards for new motor vehicles on April 1st 2010, the agency is now proceeding to control GHG emissions from new and modified stationary sources as well, including power plants and manufacturing facilities. Stationary sources account for 69% of U.S. emissions of greenhouse gases. This book provides background on stationary sources of greenhouse gas pollution and identifying options Congress has at its disposal to address the issues.

Chapter 1- As a direct result of the Environmental Protection Agency's promulgation of an "endangerment finding" for greenhouse gas (GHG) emissions in December 2009, and its subsequent promulgation of GHG emission standards for new motor vehicles on April 1, 2010, the agency is now proceeding to control GHG emissions from new and modified *stationary* sources as well, including power plants, manufacturing facilities, and others. Stationary sources account for 69% of U.S. emissions of greenhouse gases. If the United States is to reduce its total GHG emissions, as President Obama has committed to do, it will be necessary to address these sources.

Chapter 2- Stationary sources—a term that includes power plants, petroleum refineries, manufacturing facilities, and other non-mobile sources of air pollution—are not yet subject to any greenhouse gas (GHG) emission standards issued by the EPA; but because of the Clean Air Act's wording, such stationary sources *will become subject to permit requirements* for their GHG emissions beginning on January 2, 2011. Affected units will be subject to the permitting requirements of the Prevention of Significant Deterioration (PSD) and Title V provisions. For PSD, this will include state determinations of what constitutes Best Available Control Technology (BACT) that affected facilities will be required to install. On November 10, 2010, EPA released guidance and technical information to assist state authorities in issuing permits and determining BACT.

Chapter 3- As Congress and the Administration considered new legislation to reduce the greenhouse gas (GHG) emissions that contribute to climate change over the last year and a half (a process that has now stalled), the Environmental Protection Agency simultaneously began to exercise its existing authority under the Clean Air Act to set standards for GHG emissions. The Administration has made clear that its preference would be for Congress to address the climate issue through new legislation. Nevertheless, it is moving forward on several fronts to define how the Clean Air Act will be used and to promulgate regulations.

Chapter 4- Although new legislation to address greenhouse gases is a leading priority of the President and many members of Congress, the ability to limit these emissions already exists under Clean Air Act authorities that Congress has enacted – a point underlined by the Supreme Court in an April 2007 decision, *Massachusetts v. EPA*. In response to the Supreme Court decision, EPA has begun the process of using this existing authority, issuing an "endangerment finding" for greenhouse gases (GHGs) December 7, 2009, and proposing GHG regulations for new motor vehicles in the September 28, 2009 *Federal Register*.

Chapter 5- Chairman Boxer, Subcommittee Chairman Carper, Ranking Member Inhofe, Subcommittee Ranking Member Vitter, and members of the Committee, thank you for inviting me to testify today to update you on EPA's efforts to mitigate the impacts of emissions from power plants. As you will recall, I last appeared before this committee to discuss these issues in July 2009, and since that time I am pleased to report that EPA has made significant progress on our regulatory efforts to address the public health and environmental effects of air pollutants from power plants. In my testimony I will discuss the status of our work on these efforts, and will provide the committee with some information on S. 2995, the Clean Air Act Amendments of 2010.

Chapter 6- Chairman Carper, Ranking Member Vitter, and members of the Subcommittee, thank you for inviting me to testify today about EPA's efforts to mitigate the impacts of emissions from power plants. During my visits with many of you during my confirmation process to be the head of EPA's Office of Air and Radiation, I appreciated the opportunity to discuss with you our shared concerns over the public health and environmental effects of air pollutants from power plants. I agree with Senator Carper that emissions of $SO_2$, $NO_x$, mercury, and other pollutants from the generation of energy is a cause for great concern, and I am grateful for his leadership on this important issue over the years. I am glad that we have begun this dialogue and I look forward to continuing to work on this issue.

In: EPA Regulation of Greenhouse Gases ISBN: 978-1-61470-729-5
Editors: Cianni Marino and Nico Costa © 2012 Nova Science Publishers, Inc.

*Chapter 1*

# EPA REGULATION OF GREENHOUSE GASES: CONGRESSIONAL RESPONSES AND OPTIONS

## *James E. McCarthy and Larry Parker*

### SUMMARY

As a direct result of the Environmental Protection Agency's promulgation of an "endangerment finding" for greenhouse gas (GHG) emissions in December 2009, and its subsequent promulgation of GHG emission standards for new motor vehicles on April 1, 2010, the agency is now proceeding to control GHG emissions from new and modified *stationary* sources as well, including power plants, manufacturing facilities, and others. Stationary sources account for 69% of U.S. emissions of greenhouse gases. If the United States is to reduce its total GHG emissions, as President Obama has committed to do, it will be necessary to address these sources.

EPA's regulations limiting GHG emissions from new cars and light trucks automatically triggered two other Clean Air Act (CAA) provisions affecting stationary sources of air pollution. First, effective January 2, 2011, new or modified major stationary sources must undergo New Source Review (NSR) with respect to their GHGs in addition to any other pollutants subject to regulation under the CAA that are emitted by the source. This review requires affected sources to install Best Available Control Technology (BACT) to address their GHG emissions. Second, major sources of GHGs (existing and new) will have to obtain permits under Title V of the CAA (or have existing permits modified to include their GHG requirements). Beyond these permitting requirements, because stationary sources, particularly coal-fired power plants, are the largest sources of greenhouse gas emissions, EPA is likely to find itself compelled to issue endangerment findings and establish emission control standards for GHG emissions under other parts of the act. For example, in December 2010, EPA reached settlement agreements with numerous parties under which it will promulgate final decisions on New Source Performance Standards (NSPS) for electric generating units by May 2012 and for petroleum refineries by November 2012.

EPA shares congressional concerns about the potential scope of these regulations, primarily because a literal reading of the act would have required as many as 6 million stationary sources to obtain permits. To avoid this result, on May 13, 2010, the agency finalized a "Tailoring Rule" that focuses its resources on the largest emitters while deciding over a six-year period what to do about smaller sources.

Many in Congress have suggested that EPA should delay taking action on any stationary sources or should be prevented from doing so. There are at least ten bills introduced in the 112th Congress that would delay or prevent EPA actions on greenhouse gas emissions. In February, the text of one bill, H.R. 153, was added to the Full-Year Continuing Appropriations Act (H.R. 1) during floor debate on a 249-177 vote. H.R. 1 passed the House, February 19, but failed in the Senate, March 9. On April 7, the House passed Representative Upton's H.R. 910, which would repeal EPA's endangerment finding, redefine "air pollutants" to exclude greenhouse gases, and prohibit EPA from promulgating any regulation to address climate change. In the Senate, similar legislation failed to pass, April 6.

This report discusses elements of this controversy, providing background on stationary sources of greenhouse gas pollution and identifying options Congress has at its disposal to address the issues, including (1) resolutions of disapproval under the Congressional Review Act; (2) freestanding legislation; (3) the use of appropriations bills as a vehicle to restrain EPA activity; and (4) amendments to the Clean Air Act, including legislation to establish a new GHG control regime.

# INTRODUCTION

On April 1, 2010, Lisa Jackson, the Environmental Protection Agency (EPA) Administrator, signed final regulations that will require auto manufacturers to limit emissions of greenhouse gases (GHGs) from new cars and light trucks.[1] These regulations have now triggered two other Clean Air Act provisions affecting *stationary* sources of air pollution such as electric power plants. First, effective January 2, 2011, new or modified major stationary sources have to undergo New Source Review (NSR) with respect to their GHG emissions in addition to any other pollutants subject to regulation under the Clean Air Act that they emit. This review requires affected sources to install Best Available Control Technology (BACT) to address their GHG emissions. Second, existing sources (in addition to new ones) have to obtain permits under Title V of the Clean Air Act (or have existing permits modified to include their GHG requirements).

## Congressional Concerns

EPA's potential regulation of GHG emissions (particularly from stationary sources) has led some in Congress to suggest that the agency delay taking action or be stopped from proceeding. In the 112th Congress, there are at least 10 bills introduced affecting EPA's greenhouse gas authority, as well as several amendments addressing the same issues.

EPA has attempted to respond to congressional concerns by clarifying the direction and schedule of its actions. However, the agency has been limited to the degree it can delineate specifics as many of the regulatory components, such as new New Source Performance Standards (NSPS) for stationary sources, are in the early stages of the rulemaking process. EPA has provided three clear responses so far to the congressional concerns outlined above:

- the first came on March 29, 2010, when the Administrator reinterpreted a 2008 memorandum concerning the effective date of the stationary source permit requirements.[2] Facing a possibility of having to begin the permitting process on April 1, 2010 (the date the new GHG standard for automobiles was finalized), the March 29 decision delayed for nine months (to January 2, 2011) the date on which EPA would consider stationary source GHGs to be subject to regulation, and thus, subject to the permitting requirements of PSD-NSR and Title V.[3]

- On May 13, 2010, the Administrator signed the GHG "Tailoring" Rule, which provided for a phasing in of Title V and PSD-NSR permitting requirements, and is discussed in detail below.
- On November 10, 2010, the EPA released a package of guidance and technical information to assist local and state permitting authorities in implementing PSD and Title V permitting for greenhouse gas emissions.[4]

EPA Administrator Lisa Jackson and the President have repeatedly expressed their preference for Congress to take the lead in designing a GHG regulatory system. However, EPA simultaneously states that, in the absence of congressional action, it must proceed to regulate GHG emissions: a 2007 Supreme Court decision (*Massachusetts v. EPA*[5]) compelled EPA to address whether GHGs are air pollutants that endanger public health and welfare, and if so to embark on a regulatory course that is prescribed by statute. Having made an affirmative decision to the endangerment question, EPA is now proceeding with regulations.

Thus, EPA and a number of Members and Senators appear to be on a collision course. EPA is proceeding to regulate emissions of GHGs under the Clean Air Act, as it maintains it must, while trying to focus those efforts on the largest emitters within a feasible timeframe. Opponents of this effort in Congress are considering various approaches to alter the agency's course.

This report discusses elements of this controversy, providing background on stationary sources of greenhouse gas pollution and identifying options Congress has if it chooses to address the issue. The report discusses four sets of options: (1) resolutions of disapproval under the Congressional Review Act; (2) freestanding legislation delaying or prohibiting EPA action; (3) the use of appropriations bills as a vehicle to restrain EPA activity; and (4) amendments to the Clean Air Act, including legislation to establish a new GHG control regime. The report considers each of these in turn, but first provides additional detail regarding the sources of GHG emissions, the requirements of the Clean Air Act, and the significance of regulating emissions from stationary sources.

# REGULATION OF STATIONARY SOURCE GHGS

When EPA finalized the regulation of greenhouse gases from new mobile sources, legal and policy drivers were activated that are leading to regulation of stationary sources as well. Stationary sources are the major sources of the country's GHG emissions. Overall, 69% of U.S. emissions of greenhouse

## EPA Regulation of Greenhouse Gases

gases come from stationary sources (the remainder come from mobile sources, primarily cars and trucks). Relatively large sources of fossil-fuel combustion and other industrial processes are responsible for more than half the country's total emissions (see Table 1). If EPA (or Congress) is to embark on a serious effort to reduce greenhouse gas emissions, stationary sources, and in particular large stationary sources, will have to be included.

The substantial amount of greenhouse gas emissions emanating from stationary source categories is even more important from a policy standpoint: reductions in greenhouse gas emissions from these sources are likely to be more timely and cost-effective than attempts to reduce emissions from the transport sector.

### Table 1. Selected U.S. Stationary Sources of Greenhouse Gases (million metric tons of $CO_2$-equivalent)

| Source | 2009 Emissions | % of Total GHGs |
|---|---|---|
| Electricity Generation ($CO_2$, $CH_4$, $N_2O$) | | |
| Coal-fired | 1756.5 | 26.5% |
| Natural gas-fired | 373.4 | 5.6% |
| Fuel oil-fired | 33.0 | 0.5% |
| **Industrial fossil-fuel combustion ($CO_2$, $CH_4$, $N_2O$)** | | |
| Mostly petroleum refineries, chemicals, primary metals, paper, food, and nonmetallic mineral products | | |
| Coal-fired | 84.0 | 1.3% |
| Natural gas-fired | 365.3 | 5.5% |
| Fuel oil-fired | 290.5 | 4.4% |
| **Industrial Processes** | | |
| Iron and steel production ($CO_2$, $CH_4$) | 42.0 | 0.6% |
| Cement production ($CO_2$) | 29.4 | 0.4% |
| Nitric acid production ($N_2O$) | 14.6 | 0.2% |
| Substitution for ozone-depleting substances (HFCs) | 119.3 | 1.8% |
| **Other** | | |
| Natural gas systems ($CO_2$, $CH_4$) | 253.4 | 3.8% |
| Landfills ($CH_4$) | 117.5 | 1.8% |
| **TOTAL** | **3478.9** | **52.4%** |

Source: EPA, DRAFT *Inventory of U.S. Greenhouse Gas Emissions and Sinks: 1990-2009*, February 15, 2011.

Two factors are driving the concerns about EPA's decisions on mobile sources spilling over to decisions on stationary sources: (1) the non-discretionary triggers within the CAA, discussed above, that impose permitting requirements on stationary sources because of the mobile source action; and (2) legal and policy linkages between mobile and stationary sources with respect to greenhouse gases that are likely to force EPA to issue additional endangerment findings and accompanying regulations on stationary sources. In particular, three potential impacts on stationary sources have raised the most concern:

- mandatory permitting requirements under the Prevention of Significant Deterioration / New Source Review (PSD-NSR) program (Sections 165-169);
- mandatory permitting requirements under Title V, the permit title of the Clean Air Act; and
- further endangerment findings that would require greenhouse gas reductions under different parts of the act,[6] particularly Section 111, New Source Performance Standards.

## Prevention of Significant Deterioration/New Source Review (PSD-NSR)

Under Sections 165-169 of the Clean Air Act, any new or modified facility emitting (or potentially emitting) over 100 or 250 tons of any regulated pollutant[7] must undergo preconstruction review and permitting, including the installation of Best Available Control Technology (BACT). State permitting agencies determine BACT on a case-by-case basis, taking into account energy, environmental, and economic impacts. BACT cannot be less stringent than the federal New Source Performance Standard, but it can be more so.[8] EPA issues guidelines to states to assist them in making BACT determinations.[9]

PSD-NSR is required for any pollutant "subject to regulation" under the Clean Air Act, a requirement that was fulfilled when the mobile source regulations EPA finalized April 1, 2010, took effect January 2, 2011. Two aspects of invoking the New Source Review provision have been raised. First, as noted above, PSD-NSR has specified thresholds for triggering its provisions: a "major emitting facility" is defined as emitting or having the potential to emit either 100 tons or 250 tons annually of a regulated pollutant (Sec. 169(1)).[10] With respect to greenhouse gases, this is a fairly low threshold. EPA concludes that at 100 tons per year, even large residential and commercial structures could be required to obtain permits. By comparison, the Waxman-Markey bill (H.R. 2454) of the 111th Congress generally used 25,000 metric tons as a regulatory threshold.

# EPA Regulation of Greenhouse Gases

The second administrative issue for PSD-NSR is the requirement that BACT be determined on a case-by-case basis. Combined with a 100-ton or 250-ton threshold, this could mean a massive increase in state determinations of BACT: the resulting increased permit activity would be at least two orders of magnitude, according to EPA.

EPA has addressed this threshold problem in the Greenhouse Gas Tailoring Rule, signed by the Administrator May 13, 2010.[11] The rule phases in the PSD-NSR requirements:

- in Step 1, from January 2, 2011, to June 30, 2011, there will be no new permitting actions due solely to GHG emissions. Only sources undertaking permitting actions anyway for other pollutants will need to address GHGs, with a threshold of 75,000 tons per year (tpy) of $CO_2$-equivalent ($CO_2$-e) for applicability;
- in Step 2, from July 1, 2011, to June 30, 2013, new sources that are not subject to major source permit requirements for any other air pollutant will require PSD-NSR and Title V permits if they have the potential to emit 100,000 tpy or more of $CO_2$-e. Modifications of sources not otherwise subject to permit requirements will have a permit threshold of 75,000 tpy;
- in Step 3, which would require a new rulemaking from EPA, the agency will consider lowering the permit threshold, but not below 50,000 tpy of $CO_2$-e, beginning July 1, 2013;
- the agency will also complete a study within five years projecting the administrative burden of requiring permits from smaller sources, considering available streamlining measures, and will solicit comment on permanent exclusion of certain sources from PSD, Title V, or both requirements in a rulemaking to be completed by April 30, 2016.

EPA estimates that under Steps 1 and 2, 1,600 new or modified sources annually will be required to obtain NSR permits for their GHG emissions. Without the Tailoring Rule, the estimate would be that 82,000 facilities would have required permits.[12]

## Title V Permits

When invoked by EPA's mobile source action, Title V requires all new and existing facilities that have the potential to emit a GHG pollutant in amounts of 100 tons per year or more to obtain permits. This size threshold is even more stringent than the above NSR requirement. If not modified, it would result in substantial numbers of smaller sources having to obtain a state permit for the first time (most larger sources already have permits because they emit other pollutants regulated under the act).

In the preamble to its Tailoring Rule, EPA estimated that more than 6 million sources would potentially be subject to Title V if the threshold remained at 100 tons per year of emissions.[13] Thus, like PSD-NSR, a major complication that Title V introduces is the potential for very small sources of greenhouse gases to need permits in order to operate (or continue operating). Furthermore, Title V requires that covered entities pay fees established by the permitting authority, and that the total fees be sufficient to cover the costs of running the permit program.

It should be noted that Title V permits are designed to help states and the EPA in enforcing a source's various Clean Air Act-related requirements; they do not impose any requirements themselves. They simply put all the affected facility's Clean Air Act requirements in one place to make enforcement more efficient. Thus, for large facilities that already have Title V permits because of their emission of other regulated pollutants, the addition of GHGs to that permit does not represent a significant additional administrative burden. It is the potential for millions of sources not currently required to have a Title V permit that would have to obtain one under GHG regulations that represents the additional burden identified here, and is the impetus for EPA's Tailoring Rule described above. As a result of the Tailoring Rule, EPA estimates that 15,500 sources annually will need to obtain Title V permits.

## Potential GHG Emission Standards Under Section 111

Because stationary sources, particularly coal-fired power plants, are a major source of greenhouse gas emissions, EPA is likely to be compelled to issue further endangerment findings under separate parts of the act, resulting in regulation of greenhouse gases from stationary sources.[14] There are numerous paths such regulation might take: in the immediate future, the most likely route to stationary source GHG regulations would be Section 111, New Source Performance Standards (NSPS).

New Source Performance Standards are emission limitations imposed on designated categories of major new (or substantially modified) stationary sources of air pollution. A new source is subject to NSPS regardless of its location or ambient air conditions. Section 111 provides authority for EPA to impose performance standards on stationary sources—directly in the case of new (or modified) sources, and through the states in the case of existing sources (Section 111(d)). The authority to impose performance standards on new and modified sources refers to any category of sources that the Administrator judges "causes, or contributes significantly to, air pollution which may reasonably be anticipated to endanger public health or welfare" (Sec. 111(b)(1)(A))—language similar to the endangerment and cause-or-contribute findings EPA promulgated for motor vehicles on December 15, 2009.

In establishing these standards, Section 111 gives EPA considerable flexibility with respect to the source categories regulated, the size of the sources regulated, the particular gases regulated, along with the timing and phasing in of regulations (Sec. 111(b)(2)). This flexibility extends to the stringency of the regulations with respect to costs, and secondary effects, such as non-air-quality, health and environmental impacts, along with energy requirements. This flexibility is encompassed within the Administrator's authority to determine what control systems she determines have been "adequately demonstrated." Standards of performance developed by the states for existing sources under Section 111(d) can be similarly flexible.

Much attention, including EPA's, has been on this path. Section 111 gives EPA authority to set NSPS for emissions of "air pollutants," a term that now has been determined to include greenhouse gases.[15] Section 111(d), which broadens the NSPS authority to state plans for existing sources of air pollutants, refers to any air pollutant that *isn't* either a criteria air pollutant under Section 108 or a toxic air pollutant under Section 112. Again, greenhouse gases would fit within the boundaries of the term.

In addition, attention will be focused on Section 111 as any potential federally determined NSPS for new sources would constitute the "floor" for state BACT determinations under PSD-NSR. Thus, as states move to implement NSR for greenhouse gases, the pressure on EPA to set the NSPS floor on those determinations may increase.

The potential schedule for Section 111 NSPS standards has been the subject of much speculation, but for two major categories of sources (electric generating units and petroleum refineries) the speculation has now ended. On December 23, 2010, EPA announced that it had reached settlement agreements with numerous parties under which it will promulgate final decisions on NSPS for electric generating units by May 2012 and for petroleum refineries by November 2012.

How quickly such standards could be applied to existing sources is an open question. EPA must first propose and promulgate guidelines, following which the states would be given time to develop implementation plans.[16] Following approval of the plans, the act envisions case-by-case determinations of emission limits, in which the states may consider, among other factors, the remaining useful life of a source in setting an emission limit. Thus, it is likely to be several years before existing power plants are subject to emission limits for GHGs.

# CONGRESSIONAL OPTIONS

As noted earlier, if Congress would like to see a different approach to GHG controls than the one on which EPA has embarked, at least four sets of options are available to change the agency's course: the Congressional Review Act; freestanding legislation; appropriations riders; and amendments to the Clean Air Act. Among the most widely discussed has been the Congressional Review Act.

## Congressional Review Act[17]

The Congressional Review Act (CRA, 5 U.S.C. §§ 801-808), enacted in 1996, establishes special congressional procedures for disapproving a broad range of regulatory rules issued by federal agencies. Before any rule covered by the act can take effect, the federal agency that promulgates the rule must submit it to Congress. If Congress passes a joint resolution disapproving the rule under procedures provided by the act, and the resolution becomes law,[18] the rule cannot take effect or continue in effect. Also, the agency may not reissue either that rule or any substantially similar one, except under authority of a subsequently enacted law.

After EPA promulgated the first of its GHG rules, the endangerment finding[19] on December 15, 2009, four identical resolutions were introduced to disapprove it under the CRA—one in the Senate (Senator Murkowski's S.J.Res. 26) and three in the House (Representative Jerry Moran's H.J.Res. 66, Representative Skelton's H.J.Res. 76, and Representative Barton's H.J.Res. 77). If enacted, these resolutions would have disapproved both the "endangerment" and "cause or contribute" findings that EPA promulgated, with the result that the findings would have "no force or effect." These resolutions garnered substantial support: the Murkowski resolution had 40 Senate cosponsors, and the identical House measures had 3, 52, and 121 cosponsors respectively.

Nevertheless, the path to enactment of such a resolution is a steep one. The Obama Administration has made the reduction of GHG emissions one of its major goals; as a result, many have concluded that legislation restricting EPA's authority to act, if passed by Congress, would encounter a presidential veto. Overriding a veto requires a two-thirds majority in both the House and Senate, and is seen by many as unlikely.

Opponents of the above resolutions noted at least two reasons for their opposition. First, a successful resolution of disapproval for the endangerment and cause-or-contribute findings would not overturn a rule that imposes regulatory controls, but rather EPA's scientific findings that are the prerequisite for any EPA regulatory action on GHGs. Such findings are under the purview of the Congressional Review Act, but a disapproval resolution would put Congress in the position of overruling a science-based conclusion resulting from a regulatory agency's review and analysis of available scientific evidence.

Second, since the endangerment and cause-or-contribute findings were made under the motor vehicle section of the act (Section 202(a)), EPA argued that a resolution of disapproval would make it impossible for the agency's GHG standards for light duty vehicles to take effect. Section 202(a) only allows the Administrator to set standards for pollutants that she finds "may reasonably be anticipated to endanger public health or welfare." Thus, the absence of an endangerment finding would remove the prerequisite to the promulgation of standards.

The light-duty-vehicle GHG standards, promulgated April 1, 2010, are not particularly controversial in and of themselves. They are the product of negotiations among nine auto manufacturers; the states of California, Michigan, and Massachusetts; the United Auto Workers; environmental groups; EPA; the Department of Transportation; and the White House. The nine auto manufacturers, including GM, Ford, and Chrysler, support them because, in their absence, states would be free to impose GHG standards themselves, leading to what auto industry spokespersons termed a "patchwork" of regulatory requirements that would be more difficult for the automakers to meet.[20]

The CRA is designed primarily to specify the procedures under which a resolution of disapproval is to be considered in the Senate. Pursuant to the Congressional Review Act, an expedited procedure for Senate consideration of a disapproval resolution may be used at any time within 60 days of Senate session after the rule in question has been published in the *Federal Register* and received by both houses of Congress. The expedited procedure provides that, if the committee to which a disapproval resolution has been referred has not reported it by 20 calendar days after the rule has been received by Congress or published in the *Federal Register*, the panel may be discharged if 30 Senators submit a petition for that purpose. The resolution is then placed on the Calendar.

Under the expedited procedure, once a disapproval resolution is on the Calendar in the Senate, a motion to proceed to consider it is in order. Several provisions of the expedited procedure protect against various potential obstacles to the Senate's ability to take up a disapproval resolution. The Senate has treated a motion to consider a disapproval resolution under the Congressional Review Act as not debatable, so that this motion cannot be filibustered through extended debate. After the Senate takes up the disapproval resolution itself, the expedited procedure of the Congressional Review Act protects the ability of the body to continue and complete that consideration. It limits debate to 10 hours and prohibits amendments.[21]

On May 24, 2010, a unanimous-consent agreement was reached providing for a vote on S.J.Res. 26 under procedures similar to those provided by the Congressional Review Act, but on June 10, the Senate voted 47-53 not to take up the resolution.

The Congressional Review Act sets no deadline for final congressional action on a disapproval resolution, so a resolution could theoretically be brought to the Senate floor even after the expiration of the deadline for the use of the CRA's expedited procedures. To obtain floor consideration, the bill's supporters would have to follow the Senate's normal procedures. Similarly, a resolution could reach the House floor through its ordinary procedures, that is, generally by being reported by the committee of jurisdiction (in this case, the Energy and Commerce Committee). If the committee of jurisdiction does not report a disapproval resolution submitted in the House, a resolution could still reach the floor pursuant to a special rule reported by the Committee on Rules (and adopted by the House), by a motion to suspend the rules and pass it (requiring a two-thirds vote), or by discharge of the committee (requiring a majority of the House [218 members] to sign a petition).

If either house passes a disapproval resolution, the CRA provides that the other house should consider its own companion measure, but then vote on the measure received from the house that acted first. This procedure facilitates clearing the measure for presentation to the President. Yet the CRA establishes no expedited procedure for further congressional action on a disapproval resolution if the President vetoes it. In such a case, Congress would need to attempt an override of a veto using its normal procedures for considering vetoed bills.

## *Freestanding Legislation*

To provide for a more nuanced response to the issue than permitted under the CRA, some members have introduced freestanding legislation. At least 10 bills (and several amendments) have been introduced in the 112th Congress that would prohibit temporarily or permanently EPA's regulation of greenhouse gas emissions. These bills face the same obstacle as a CRA resolution of disapproval, however (i.e., being subject to a presidential veto). Among those introduced, attention has focused on two: Representative Upton's and Senator Inhofe's H.R. 910 / S. 482, and Senator Rockefeller's and Representative Capito's S. 231 / H.R. 199.

## *S. 231/H.R. 199*

S. 231, introduced by Senator Rockefeller, and H.R. 199, introduced by Representative Capito, are similar bills that basically reintroduce Senator Rockefeller's bill of the 111th Congress. S. 231, entitled the EPA Stationary Source Regulations Suspension Act and H.R. 199, entitled the Protect America's Energy and Manufacturing Jobs Act of 2011, provide that during the two-year period beginning on the date of their enactment, EPA could not take any action under the Clean Air Act with respect to any stationary source permitting requirement or any requirement under the New Source Performance Standards section of the act relating to carbon dioxide or methane.[22] A stated reason for the two-year delay would be to allow Congress to enact legislation specifically designed to address climate change. By specifically identifying stationary sources and the two specific pollutants as its objectives, the bill would allow EPA to proceed with GHG controls for mobile sources (including, cars, trucks, ships, aircraft, and nonroad engines of all kinds— which account for 31% of U.S. greenhouse gas emissions), and it would allow the agency to regulate emissions of non-$CO_2$ and non-methane greenhouse gases (including nitrous oxide, hydrofluorocarbons, perfluorocarbons, and sulfur hexafluoride, which together account for 6.8% of GHGs, expressed as $CO_2$-equivalents). The bill was designed to be more acceptable to members willing to consider a delay of EPA action, as opposed to overturning EPA's scientific conclusions or blocking EPA action altogether. The bill was offered

14 James E. McCarthy and Larry Parker

as an amendment to S. 493 (S.Amdt. 215) on April 6, and was not agreed to, on a vote of 12-88.

### *H.R. 910/S. 482/S.Amdt. 183*

Chairman Upton of the House Energy and Commerce Committee and Senator Inhofe, ranking member of the Senate Environment and Public Works Committee, have introduced legislation to permanently remove EPA's authority to regulate greenhouse gases (H.R. 910/S. 482). The House version was reported (amended) by the Energy and Commerce Committee April 1, 2011, and was passed by the House, 255-177, April 7. In the Senate, Senator McConnell introduced language identical to Senator Inhofe's bill as an amendment to S. 493 (S.Amdt. 183). The amendment was not agreed to, on a vote of 50-50, April 6. The Upton-Inhofe-McConnell bill's provisions are similar in many respects to a bill introduced by Senator Barrasso, S. 228. Like S. 228, the bill would repeal a dozen EPA greenhouse-gas-related regulations, including the Mandatory Greenhouse Gas Reporting rule, the Endangerment Finding, and the PSD and Title V permitting requirements. It would redefine the term "air pollutant" to exclude greenhouse gases. And it states that EPA may not "promulgate any regulation concerning, take action related to, or take into consideration the emission of a greenhouse gas to address climate change."

The bill would have no effect on Title VI of the Clean Air Act (ozone depletion), or federal research, development, and demonstration programs. The current light-duty motor vehicle GHG standards and the proposed GHG emission standards for Medium- and Heavy-Duty Engines and Vehicles would be allowed to continue, but no future ones would be allowed.

The bill would not pre-empt state authority to regulate greenhouse gases, but would not allow EPA to permit such regulations within a state SIP or to federally enforce them. Also, EPA would be prohibited from granting another California waiver for greenhouse gas controls from mobile sources.

### *Other Amendments to S. 493*

In addition to the McConnell and Rockefeller amendments, two other amendments to S. 493 to address EPA's greenhouse gas authority were considered in the Senate on April 6. One was Senator Baucus's S.Amdt. 236; the other was S.Amdt. 277, authored by Senator Stabenow and Senator Sherrod Brown. S.Amdt. 236 would have set thresholds (similar to EPA's "Tailoring Rule") to exempt most sources of greenhouse gas emissions from having to obtain Clean Air Act permits for those emissions. It would also have excluded agricultural sources from PSD-NSR permitting requirements based on their GHG emissions. The Stabenow-Brown amendment would have suspended EPA greenhouse gas requirements for stationary sources, including permits and New Source Performance Standards, for a two-year period. It would have exempted GHG emissions from agricultural sources from regulation. And it would have extended the tax credit for Advanced Energy Projects, with an authorization of $5 billion. Both the Baucus and Stabenow-Brown amendments were not agreed to, on votes of 7-93.

## Appropriations Bills

A third option that Congress has used to delay regulatory initiatives is to place an amendment, or "rider" on the agency's appropriation bill that prevents funds from being used for the targeted initiative. In its FY2011 budget submission,[23] EPA requested $43 million for "additional regulatory efforts aimed at taking action on climate change," $25 million "for state grants focused on developing technical capacity to address greenhouse gas emissions under the Clean Air Act," and $13.5 million "for implementing new emission standards that will reduce Greenhouse Gas (GHG) emissions from mobile sources" including "developing potential standards for large transportation sources such as locomotives and aircraft engines, and analyzing the potential need for standards under petitions relating to major stationary sources."[24] These are small sums in an agency budget request of slightly more than $10 billion, but GHG regulations have been among the most controversial questions at congressional hearings on the agency's budget submission. Thus, it was not surprising to see further discussion of amendments to the EPA appropriation or report language limiting or delaying EPA's GHG regulatory actions.

In comparison to a CRA resolution of disapproval or stand-alone legislation, addressing the issue through an amendment to the EPA appropriation—an approach that was discussed at some length in the fall of 2009, when Senator Murkowski and others drafted but ultimately did not introduce amendments to the FY2010 Interior Appropriation—may be considered easier. The overall appropriation bill to which it would be attached would presumably contain other elements that would make it more difficult to veto.

Funding amendments might take several forms. Since it is the triggering of standards for stationary sources (power plants, manufacturing facilities, and others) that has raised the most concern, however, it seems fair to assume that any effort to delay or prevent EPA action under an appropriations rider, like the freestanding legislation discussed above, would focus on these sources. An example of this focus is Representative Poe's H.R. 153. H.R. 153 would prohibit EPA funding for implementing or enforcing a greenhouse gas cap-and-trade program or any other greenhouse gas regulatory requirement on stationary sources issued or effective after January 1, 2011 (including permitting requirements under PSD and Title V). This bill would permit continued regulation of mobile sources as it only affects stationary sources.

FY2011 appropriations for EPA and the rest of the government were provided through early April, 2011, by a series of continuing resolutions, leaving the question of EPA appropriations and potential riders affecting the agency's GHG regulatory efforts for the 112th Congress to decide. In February, language similar to H.R. 153 was added to the Full-Year Continuing Appropriations Act, 2011 bill (H.R. 1) during floor debate on a 249-177 vote (H.Amdt. 101). However, the Senate failed to pass the bill, 44-56, March 9.

## Amending the Clean Air Act

The most comprehensive approach that Congress might take to alter EPA's course would be to amend the Clean Air Act to modify EPA's current regulatory authority as it pertains to GHGs. This was the option chosen by the House in passing H.R. 2454, the American Clean Energy and Security Act (the Waxman-Markey bill) and by the Senate Environment and Public Works Committee in its reporting of S. 1733, the Clean Energy Jobs and American Power Act (the Kerry-Boxer bill) in the 111th Congress. The bills would have amended the Clean Air Act to establish an economy-wide cap-and-trade program for GHGs, established a separate cap-and-trade program for HFCs, preserved EPA's authority to regulate GHG emissions from mobile sources while setting deadlines for regulating specific mobile source categories, and required the setting of New Source Performance Standards for uncapped major sources of GHGs.

At the same time, both bills contained provisions to limit EPA's authority to set GHG standards or regulate GHG emissions under Sections 108 (National Ambient Air Quality Standards), 112 (Hazardous Air Pollutants), 115 (International Air Pollution), 165 (PSD-NSR), and Title V (Permits) because of the climate effects of these pollutants.[25] The bills would not have prevented EPA from acting under these authorities if one or more of these gases proved to have effects other than climate effects that endanger public health or welfare.

With respect to exemption from the permitting requirements of the PSD program and Title V, the bills differed in the extent of their exemptions. The H.R. 2454 provision would have prevented new or modified stationary sources from coming under the PSD-NSR program solely because they emit GHGs. In contrast, the Senate bill's provision would have simply raised the threshold for regulation under PSD from the current 100 or 250 short tons to 25,000 metric tons with respect to any GHG, or combination of GHGs. Likewise, with respect to Title V permitting, the H.R. 2454 provision would have prevented any source (large or small) from having to obtain a state permit under Title V solely because they emit GHGs. In contrast, the exemption under the Senate bill was restricted to sources that emit under 25,000 metric tons of any GHG or combination of GHGs.[26]

Amending the Clean Air Act to revoke some existing regulatory authority as it pertains to GHGs while establishing new authority designed specifically to address their emissions is the approach advocated by the Administration and, indeed, by many participants in the climate debate regardless of their position on EPA's regulatory initiatives. However, the specifics of a bill acceptable to a majority would be difficult to craft.

## CONCLUSION

In some respects, EPA's greenhouse gas decisions are similar to actions it has taken previously for other pollutants. Beginning in 1970, and reaffirmed by amendments in 1977 and 1990, Congress gave the agency broad authority to identify pollutants and to proceed with regulation. Congress did not itself identify the pollutants to be covered by National Ambient Air Quality Standards (NAAQS), for example; rather, it told the agency to identify pollutants that are emitted by numerous and diverse sources, and the presence of which in ambient air endangers public health and welfare. EPA has used this authority to regulate six pollutants or groups of pollutants, the so-called "criteria pollutants."[27] EPA also has authority under other sections of the act— notably Sections 111 (New Source Performance Standards), 112 (Hazardous Air Pollutants), and 202 (Motor Vehicle Emission Standards)—to identify pollutants on its own initiative and promulgate emission standards for them.

Actions with regard to GHGs follow these precedents and can use the same statutory authorities. The differences are of scale and of degree.

Greenhouse gases are global pollutants to a greater extent than most of the pollutants previously regulated under the act:[28] Reductions in U.S. emissions without simultaneous reductions by other countries may somewhat diminish but will not solve the problems the emissions cause.[29] Also, GHGs are such pervasive pollutants, and arise from so many sources, that reducing the emissions may have broader effects on the economy than most previous EPA regulations.

EPA's focus on Section 111 as the most likely vehicle for controlling GHGs from stationary sources may reflect concerns both about potential economic effects and about implementation difficulties with respect to controlling such pervasive pollutants. Indeed, in a 2008 *Federal Register* notice, EPA made an argument that authority for a market-based control program may exist under Section 111.[30] Even if that argument fails to pass legal scrutiny, the section does provide EPA with substantial authority to address economic and implementation issues in tailoring its GHG response to the various realities surrounding stationary source controls.

Nevertheless, as noted here, the Administration's position has been that a new market-based program authorized by new legislation is the preferred option for controlling GHGs. New legislation is also the preferred option of many in Congress, regardless of whether they agree or disagree with EPA's regulatory initiatives. Until the issue is resolved through legislative negotiations or through legal or regulatory venues, EPA will likely proceed under existing authorities of the Clean Air Act and the complex interplay of legal, regulatory, and legislative events will continue.

## End Notes

[1] The regulations, which take effect with the 2012 model year, appeared in the *Federal Register* on May 7, 2010, at 75 *Federal Register* 25324. Related information is available on EPA's website at http://www.epa.gov/otaq/climate/ regulations.htm.

[2] The reinterpretation memo appeared in the *Federal Register*, April 2, 2010, at 75 *Federal Register* 17004.

[3] The term "subject to regulation" is the key Clean Air Act term that determines when affected sources would be subject to the permitting requirements of NSR and Title V. By interpreting the term to refer to January 2, 2011, rather than the date of the final regulations implementing the mobile source endangerment finding (April 1, 2010), EPA effectively delayed the impact of that rulemaking on stationary sources for nine months. For a further discussion of the term, "subject to regulation," see CRS Report R40984, *Legal Consequences of EPA's Endangerment Finding for New Motor Vehicle Greenhouse Gas Emissions*, by Robert Meltz.

## EPA Regulation of Greenhouse Gases 19

[4] U.S. EPA, Office of Air And Radiation, "PSD and Title V Permitting Guidance for Greenhouse Gases," November 2010 (subsequently revised, March 2011), at http://www.epa.gov/nsr/ghgdocs/ghgpermittingguidance.pdf.

[5] 549 U.S. 497 (2007). For more information, see CRS Report R41505, *EPA's BACT Guidance for Greenhouse Gases from Stationary Sources*, by Larry Parker and James E. McCarthy.

[6] For a further discussion of the act's various endangerment finding provisions, see CRS Report R40984, *Legal Consequences of EPA's Endangerment Finding for New Motor Vehicle Greenhouse Gas Emissions*, by Robert Meltz.

[7] Except those pollutants regulated under Sections 112 (hazardous air pollutants) and 211(o) (renewable fuels).

[8] The PSD program (Part C of Title I of the CAA) focuses on ambient concentrations of sulfur dioxide ($SO_2$), nitrogen oxides (NOx), and particulate matter (PM) in "clean" air areas of the country (i.e., areas where air quality is better than the air quality standards (NAAQS)). The program allows some increase in clean areas' pollution concentrations depending on their classification. In general, historic or recreation areas (e.g., national parks) are classified Class I with very little degradation allowed, while most other areas are classified Class II with moderate degradation allowed. States are allowed to reclassify Class II areas to Class III areas, which would be permitted to degrade up to the NAAQS, but none have ever been reclassified to Class III. There are no PSD emission limitations for GHGs, nor is there a NAAQS for GHGs. This presumably gives EPA and the states increased latitude in determining how much additional GHG pollution can be allowed by a new or modified source.

[9] See CRS Report R41505, *EPA's BACT Guidance for Greenhouse Gases from Stationary Sources*, by Larry Parker and James E. McCarthy.

[10] Section 169(1) lists 28 categories of sources for which the threshold is to be 100 tons of emissions per year. For all other sources, the threshold is 250 tons. It should be noted that a different threshold applies in the case of major modifications, which are defined by regulation, not statute. For sulfur dioxide and nitrogen oxides, the threshold for a major modification is an increase in emissions of 40 tons per year. Facilities exceeding that threshold are subject to NSR.

Given that EPA has identified by regulation the *de minimis* emission increases for triggering NSR review for modifications, it is possible EPA could set a substantially higher level for at least carbon dioxide emissions, and perhaps other greenhouse gases, if it determined such thresholds were appropriate. In the final Tailoring Rule, the agency set a threshold of 75,000 tons per year of $CO_2$-equivalent for applying NSR to modifications.

[11] The rule appeared in the June 3 *Federal Register*. See U.S. EPA, "Prevention of Significant Deterioration and Title V Greenhouse Gas Tailoring Rule," 75 *Federal Register* 31514. A six-page EPA Fact Sheet summarizing the rule is available at http://www.epa.gov/nsr/documents/20100413fs.pdf.

[12] U.S. EPA, Office of Air Quality Planning and Standards, "Summary of Clean Air Act Permitting Burdens With and Without the Tailoring Rule," p. 6, at http://www.epa.gov/nsr/documents/20100413piecharts.pdf.

[13] 75 *Federal Register* 31547, Table VI-1, p. 31547. All but 3% of these sources would be commercial establishments and large residences, according to EPA.

[14] For a discussion of the similarities and differences in the various endangerment findings contained in the Clean Air Act, see CRS Report R40984, *Legal Consequences of EPA's Endangerment Finding for New Motor Vehicle Greenhouse Gas Emissions*, by Robert Meltz.

[15] GHGs would likely be considered a "designated pollutant" under Section 111. The term "designated pollutant" is a catch-all phrase for any air pollutant that isn't either a criteria air pollutant under Section 108 or a toxic air pollutant under Section 112. Examples of these include fluorides from phosphate fertilizer manufacturing or primary aluminum reduction, or sulfuric acid mist from sulfuric acid plants.

[16] How much time the states would be given to submit plans is unclear. The statute says that the regulations shall establish a procedure "similar to that" provided for State Implementation Plans under Section 110, which generally give states three years to submit a plan, following which EPA reviews it to determine its adequacy.

[17] This section of this report, discussing the effect of the Congressional Review Act, the procedures under which a disapproval resolution is taken up in the Senate, floor consideration in the Senate, and final congressional action, is adapted from CRS Report RL31160, *Disapproval of Regulations by Congress: Procedure Under the Congressional Review Act*, by Richard S. Beth. Additional discussion of the form of disapproval resolutions, statutory time frames, other elements of the expedited procedures, and limitations of the expedited procedures can be found in that report.

[18] For the resolution to become law, the President must sign it or allow it to become law without his signature, or the Congress must override a presidential veto.

[19] 74 *Federal Register* 66496. While generally referred to as the "endangerment finding" (singular), the *Federal Register* notice consists of two separate findings: a Finding that Emissions of Greenhouse Gases Endanger Public Health and Welfare, and a Finding that Greenhouse Gases From Motor Vehicles Cause or Contribute to the Endangerment of Public Health and Welfare.

[20] For additional information on the motor vehicle standards, see CRS Report R40166, *Automobile and Light Truck Fuel Economy: The CAFE Standards*, by Brent D. Yacobucci and Robert Bamberger.

[21] These provisions help to ensure that the Senate disapproval resolution will remain identical, at least in substantive effect, to the House joint resolution disapproving the same rule, so that no filibuster is possible on the resolution itself. In addition, once the motion to proceed is adopted, the resolution becomes "the unfinished business of the Senate until disposed of," and a non-debatable motion may be offered to limit the time for debate further. Finally, the act provides that at the conclusion of debate, the Senate automatically proceeds to vote on the resolution.

[22] The phrase "relating to carbon dioxide or methane," presumably modifies both the permitting and regulation-setting prohibitions.

[23] EPA's appropriations are part of the Interior, Environment, and Related Agencies appropriation.

[24] Testimony of Lisa P. Jackson, Administrator, U.S. Environmental Protection Agency, "Hearing on the President's Proposed EPA Budget for FY 2011," Senate Environment and Public Works Committee, February 23, 2010, pp. 2-3.

[25] The Clean Air Act exemption provisions under H.R. 2454 were in Part C, Sections 831-835; under S. 1733, the provisions were in Section 128(g).

[26] For further information, see CRS Report R40896, *Climate Change: Comparison of the Cap-and-Trade Provisions in H.R. 2454 and S. 1733*, by Brent D. Yacobucci, Jonathan L. Ramseur, and Larry Parker.

[27] The six are ozone, particulate matter, carbon monoxide, sulfur dioxide, nitrogen dioxide, and lead.

[28] An exception would be chlorofluorocarbons, regulated under Title VI of the act to protect the stratospheric ozone layer. This also was a global problem, but in this case an international agreement, the Montreal Protocol, preceded EPA action and the enactment of Clean Air Act authority.

[29] However, the Administration is working in parallel internationally to obtain commitments to global GHG reductions. Demonstrating timely and significant progress toward reduction of U.S. GHG emissions is considered essential by most experts for success internationally.

[30] U.S. Environmental Protection Agency, "Regulating Greenhouse Gas Emissions Under the Clean Air Act; Proposed Rule," 73 *Federal Register* 44514-44516, July 30, 2008. Whether EPA can set up a cap-and-trade program under the Clean Air Act is the subject of considerable debate in the literature. See Lisa Heinzerling, Testimony Before the

Subcommittee on Energy and Air Quality of the Committee on Energy and Commerce, Hearing (April 10, 2008); Robert R. Nordhaus, "New Wine into Old Bottles: The Feasibility of Greenhouse Gas Regulation Under the Clean Air Act," *N.Y.U. Environmental Law Journal* (2007), pp. 53-72; Inimai M. Chettiar and Jason A. Schwartz, *The Road Ahead: EPA's Options and Obligations For Regulating Greenhouse Gases* (April 2009); and Alaine Ginocchio, et al., *The Boundaries of Executive Authority: Using Executive Orders to Implement Federal Climate Change Policy* (February 2008).

In: EPA Regulation of Greenhouse Gases      ISBN: 978-1-61470-729-5
Editors: Cianni Marino and Nico Costa    © 2012Nova Science Publishers, Inc.

*Chapter 2*

# EPA'S BACT GUIDANCE FOR GREENHOUSE GASES FROM STATIONARY SOURCES

## *Larry Parker and James E. McCarthy*

### SUMMARY

Stationary sources—a term that includes power plants, petroleum refineries, manufacturing facilities, and other non-mobile sources of air pollution—are not yet subject to any greenhouse gas (GHG) emission standards issued by the EPA; but because of the Clean Air Act's wording, such stationary sources *will become subject to permit requirements* for their GHG emissions beginning on January 2, 2011. Affected units will be subject to the permitting requirements of the Prevention of Significant Deterioration (PSD) and Title V provisions. For PSD, this will include state determinations of what constitutes Best Available Control Technology (BACT) that affected facilities will be required to install. On November 10, 2010, EPA released guidance and technical information to assist state authorities in issuing permits and determining BACT.

Among the sources likely to be affected by implementation of the PSD permit requirements are new and modified electric generating units of all kinds, but particularly those fired by coal. These sources emit substantially more than EPA's threshold of 100,000 metric tons of $CO_2$ annually: for example, a 500 megawatt (MW) coal-fired baseload power plant would emit on the order of three million metric tons of $CO_2$ annually. The coal mining industry and coal-fired electric utilities face at least half a dozen major regulatory actions over the next few years; industry supporters view these rules collectively as a significant threat to the future of coal. Viewed in this context, the permit requirement is one more nail in what increasingly appears to them as coal's future coffin.

In its new guidance, EPA retains the basic five-step process for determining BACT that it has recommended to state authorities for 20 years. The primary foci of the EPA guidance package are on state discretion in determining BACT and on energy efficiency as the most likely result of a GHG BACT analysis. These foci are evident through EPA's guidance for each of the five steps.

For those looking for bright lines and specific recommendations with respect to GHG BACT technologies, particularly with respect to coal-fired facilities, the released package does not provide them. Indeed, EPA's supplemental "Questions and Answers" release on the guidance seems to stress that it did not draw such conclusions. For example:

- Do these tools identify BACT for specific types of industrial facilities? No.
- Does this guidance say that fuel switching (coal to natural gas) should be selected as BACT for a power plant? No.
- Does this guidance say that carbon capture and storage (CCS) should be selected as BACT? No.

Likewise, the guidance provides no cost thresholds for permitting authorities to consider in determining the economic impacts of alternatives nor proposes a new approach to selecting BACT for GHG emissions. Instead, the guidance focuses on the discretionary authority that states have in determining BACT—discretion that ensures that BACT will continue to be determined on a case-by-case basis with states differing in what they consider appropriate control measures and what constitutes BACT. Whether industry will find such discretion provides sufficient regulatory certainty to invest billions in new plants remains to be seen.

In short, the EPA GHG guidance is a simple expansion of the five-step BACT process that has been used for two decades to include greenhouse gases. Whether that is an adequate response will be determined by applicants, state authorities, and future EPA regulatory actions under related parts of the act, such as Section 111 (NSPS), to which BACT is linked.

# Introduction

Over the past year, there has been increasing congressional interest in steps being taken by the Environmental Protection Agency (EPA) to address emissions of greenhouse gases (GHGs).[1] During that time, EPA has promulgated rules to (1) require reporting of GHG emissions by stationary sources that emit 25,000 tons or more of carbon dioxide equivalents ($CO_2$e); (2) set GHG emission standards for light duty motor vehicles (cars, minivans, SUVs, and light trucks); and (3) address several issues related to permit requirements for stationary sources of GHGs.[2]

Stationary sources—a term that includes electric power plants, petroleum refineries, manufacturing facilities, and other non-mobile sources of air pollution—have not yet been subject to any GHG emission standards issued by EPA; but because of the Clean Air Act's wording, such stationary sources *will become subject to permit requirements* for their GHG emissions beginning on January 2, 2011.

Among the sources likely to be affected by implementation of the Prevention of Significant Deterioration (P SD) permit requirements are new and modified electric generating units of all kinds, but particularly those fired by coal. These sources emit substantially more than EPA's threshold of 100,000 metric tons of $CO_2$ annually: for example, a 500 megawatt (MW) coal-fired baseload power plant would emit on the order of 3 million metric tons of $CO_2$ annually. The coal mining industry and coal-fired electric utilities face at least half a dozen major regulatory actions over the next few years;[3] industry supporters view these rules collectively as a significant threat to the future of coal. Viewed in this context, the PSD permit requirement is one more nail in what increasingly appears to them as coal's future coffin.

Of particular concern is Section 165's requirement that PSD permits require new and modified major sources to install the Best Available Control Technology (BACT) for each pollutant subject to regulation under the act.

This report reviews the development of EPA's November 10, 2010 GHG BACT guidance and discusses the elements of the guidance, with particular attention to their potential impact on coal- fired electric generating units.

# BEST AVAILABLE CONTROL TECHNOLOGY

## What is PSD-BACT?

"the proposed facility is subject to the best available control technology for each pollutant subject to regulation under this Act emitted from, or which results from, such facility." (Section 1 65(a)(4))

Under Sections 165-169 of the Clean Air Act, any new or modified facility emitting (or potentially emitting) over 100 or 250 tons[4] of any regulated pollutant[5] must undergo preconstruction review and permitting. This process, called New Source Review (NSR), has four major requirements:

- *Best Available Control Technology (BACT)*—determining the emissions limitation that will achieve the maximum degree of emissions reductions through application of production processes and available methods, systems, and techniques, taking into consideration energy, environmental and economic impacts.
- *Air Quality Analysis*—assessing existing ambient concentrations of air pollutants and modeling anticipated pollutant concentrations resulting from the proposed or modified project and from future growth associated with the project.
- *Impact Analysis*—assessing the air, ground, and water pollution impacts on soils, vegetation, and visibility from pollutant increases projected to result from the proposed or modified project.
- *Public Involvement*—providing opportunities for public involvement during the permit process, including hearings and appeals.

BACT is determined by state permitting agencies on a case-by-case basis, taking into account a proposed control measure's energy, environmental, and economic impacts. BACT cannot be less stringent than the federal New Source Performance Standards (NSPS), but it can be more so.[6] EPA issues guidance to states to assist them in making BACT determinations.

Greenhouse gases will be subject to regulation beginning on January 2, 2011, when the emission standards for light duty motor vehicles take effect. Thus, as of that date, new and modified major stationary sources of greenhouse gases will require permits for construction and operation.

## Who Will Have to Get a GHG Permit?

As noted above, major stationary sources are defined in the statute as those that have the potential to emit more than 100 tons or 250 tons of pollutants subject to regulation annually. For greenhouse gases, this is a relatively low threshold: EPA has estimated that more than six million existing stationary sources emit 100 tons or more of GHGs annually. The agency estimates that currently about 280 sources have required PSD permits annually, and that a 100/250 ton threshold for GHGs would increase that number by 140-fold.[7] EPA argues that expanding the PSD permit program 140-fold would pose an extraordinary burden on itself and state permitting authorities, in addition to the burden it would pose for the regulated emission sources.

The agency has sought to limit this burden through a priority-setting regulation called the "Tailoring Rule."[8] (The rule would limit the reach of both the PSD permitting program as well as the broader permit requirement for *existing* sources of GHGs under Title V of the act.) Faced with what it considers the "absurd results" of following the letter of the law, EPA in September 2009 proposed that the agency and state permitting authorities would, out of "administrative necessity," focus first on the largest facilities. As proposed, the Tailoring Rule would initially have limited the PSD and Title V permit requirements to facilities that emit more than 25,000 metric tons per year of carbon dioxide (or its equivalent in other GHGs). In the final version of the rule, the threshold was increased to 75,000 metric tons for sources that were otherwise subject to permit requirements, or 100,000 tons if the source's emissions of other pollutants would not be sufficient to require a permit.

# WHAT IS EPA'S ROLE IN DETERMINING BACT?

To assist states in making BACT determinations, EPA has provided a recommended procedure for states to use and guidance with respect to acceptable methodologies and requirements in making a determination. In addition, EPA reviews state determinations for BACT and hears appeals from parties dissatisfied with the state's determination.

The EPA procedure for determining BACT (required for federally run programs, encouraged for EPA-approved, state-run programs) is a fairly straightforward "top-down" process.[9] The overall presumption of the process is that the measure that results in the maximum reduction in the pollutant should be installed unless energy, environmental, and economic impacts of that choice justify its rejection. The five-step process, as used by EPA is as follows:

1.  *Identify Available Control Options*—there are at least five categories of control options that states can review in identifying possible BACT in an individual case:[10] (a) existing control technologies for sources of that type; (b) technically feasible options that are used on other source categories, but not the one under review; (c) innovative control technology that has never been commercially demonstrated (*inclusion not required*); (d) inherently lower polluting production processes, fuels, and coatings that can be evaluated alone or in combination with other control devices; and (e) specific design or operational parameters that may include such factors as combustion control techniques. These categories of options are not mutually exclusive. As stated by EPA in its guidance: "Combinations of inherently lower-polluting processes/practices (or a process made to be inherently less polluting) and add-on controls are likely to yield more effective means of emission control than either approach alone."[11] Data for control options include EPA's BACT/LAER[12] Clearinghouse, existing EPA or state permits, equipment vendors, trade associations, permitting engineers, and technical papers and journals.
2.  *Eliminate Technically Infeasible Control Options*—control options need to be either demonstrated on a like facility or determined to be both available and applicable in the particular case. If not, the option is eliminated from the list.
3.  *Rank Remaining Options Based on Pollutant Reduction*—a variety of performance metrics may be necessary to determine comparative control efficiencies among different options.
4.  *Eliminate Options that Fail Energy, Environmental, or Economic Criteria*—the permitting agency has discretion in weighting the three statutory criteria for exclusion.
5.  *Determine BACT*—the most effective option remaining after the steps above have been taken is determined to be BACT and the permitting agency establishes a corresponding emissions limit.

A substantial record is built during the NSR process. A completed permit application is reviewed not only by the permitting agency, but also by the Regional EPA office, Federal Land Managers (if a PSD Class I area is involved), and by the public. Conflicts among these parties and the applicant can send the permit application through a series of state and federal administrative appeals processes, along with state and federal litigation. The specifics on this process vary from state to state.

EPA provides guidance to the states in determining BACT by identifying appropriate methodologies and requirements to assist the states in identifying all potential BACT options and in eliminating options that don't meet statutory criteria.

## WHAT ARE THE OPTIONS FOR COAL-FIRED POWER PLANTS?

Much of the discussion about EPA's guidance for GHG BACT has focused on coal-fired power plants. Coal-fired electric generating facilities are responsible for about 28% of the country's total greenhouse gas emissions. As noted, in general, there are five categories of control options that are considered in a BACT analysis:[13] A review of the categories of GHG control options for coal- fired facilities reveals few readily available alternatives to significantly reduce emissions.

1.  *Existing Control Technology*: a technology proven to work for the particular source category being permitted. There are no existing add-on technologies to reduce carbon dioxide emissions from coal-fired facilities.
2.  *Technically Feasible Technology*: a technology proven to work for other source categories, but not demonstrated on the particular source category being permitted. While an argument could be made that capture technology is commercially available for industrial facilities (to separate carbon dioxide from a facility's other emissions), there is no such available technology with respect to the storage of any captured carbon dioxide. It is possible in some cases that the carbon dioxide could be piped to a facility needing carbon dioxide (such as an oil field using enhanced oil recovery (EOR)), but those opportunities would be very site-specific.
3.  *Innovative Control Technology*: a technology that has never been commercialized on any source category; it is in the pilot-plant or demonstration stage of development or deployment (for example, oxygen combustion [also called oxyfuel]). Deployment of an innovative technology on a coal-fired power plant would likely involve substantial

costs and risks—costs and risks that could jeopardize the facility's viability.

4. *Inherently Lower Polluting Production Processes, Fuels, and Coatings*: an individual adjustment or combination of adjustments in a facility's production process, its source of fuel, and its use of coatings. Inherently lower emitting fuels are available that could either partially (co-fired) or completely replace coal. Two such alternatives would be natural gas and biomass. However, these alternatives could arguably change (or "redefine") the fundamental purpose of the source (a change EPA does not require under its BACT guidance), and eliminate some of the advantages of building a coal-fired facility.

5. *Specific Design or Operational Parameters*: adjustments in a facility's design and/or inclusion of inherently less polluting work practices in the facility's operation. For a coal-fired facility, the BACT analysis would probably focus on this last category—specifically, either changing the coal-fired design to better accommodate future technology advancements (such as using integrated gasification combined cycle technology) or employing energy-efficient control measures, as discussed later.

With a multi-billion dollar investment in a project designed to operate for decades at stake, the lack of a definitive option arguably adds more uncertainty to future coal-fired power plants.

# WHAT WORK HAS EPA ALREADY DONE ON BACT?

With the options currently limited with respect to BACT for coal-fired power plants, much of EPA's work has focused on energy efficiency and innovative control measures, as discussed below.

## Bush Administration's ANPR

In April 2007, in a case involving a petition to EPA to establish GHG emission standards for motor vehicles, the Supreme Court ruled that EPA must address the question of whether GHGs cause or contribute to air pollution that endangers public health or welfare. The case was Massachusetts v. EPA. [14] For nearly two years following the Supreme Court's decision the Bush Administration's EPA did not respond to the original petition nor make a finding regarding endangerment. Its only formal action following the Court decision was to issue a detailed information request, called an Advance Notice of Proposed Rulemaking (ANPR), on July 30, 2008.[15]

The ANPR occupied 167 pages of the *Federal Register*. Besides requesting information, it took the unusual approach of presenting statements from the Office of Management and Budget, four Cabinet Departments (Agriculture, Commerce, Transportation, and Energy), the Chairman of the Council on Environmental Quality, the Director of the President's Office of Science and Technology Policy, the Chairman of the Council of Economic Advisers, and the Chief Counsel for Advocacy at the Small Business Administration, each of whom expressed their objections to regulating greenhouse gas emissions under the Clean Air Act. The OMB statement began by noting that, "The issues raised during interagency review are so significant that we have been unable to reach interagency consensus in a timely way, and as a result, this staff draft cannot be considered Administration policy or representative of the views of the Administration."[16] It went on to state that "... the Clean Air Act is a deeply flawed and unsuitable vehicle for reducing greenhouse gas emissions."[17] The other letters concurred.

In its Technical Support Document for its ANPR, EPA took a narrow view of the alternatives available to it in imposing greenhouse gas performance standards.[18] For existing electric generating sources, the EPA focused on incremental improvements in the heat rates of existing units through options that "are well known in the industry" with an overall improvement in efficiency likely to be less than 5%. For new electric generating sources, EPA noted the availability of more efficient supercritical coal units, the future availability of ultra-supercritical units, and the possibility of limited biomass co-firing.

Continuing along this line of reasoning, EPA also suggested that it could develop regulations that anticipate future technology. For example, a phase-in approach to applying $CO_2$ standards to power plants would be to mandate that "carbon-ready" generating technology be required for new construction. The objective would be to anticipate the widespread need for some form of carbon capture technology in the future by preparing for it with compatible fossil-fuel combustion technology now. The technology most discussed is integrated-gasification, combined-cycle (IGCC). With respect to some of the carbon capture technology under development, IGCC has certain advantages over pulverized coal technology. However, just how much IGCC is "carbon ready" is subject to debate. EPA stated in its ANPR that it believes such a staged approach is available to it under section 111 (the statutory floor for BACT determinations, and a possibility that EPA is also exploring for its BACT guidance):

> EPA believes that section 111 may be used to set both single-phase performance standards based upon current technology and to set two-phased or multi-phased standards with more stringent limits in future years. Future-year limits may permissibly be based on technologies that, at the time of the rulemaking, we find adequately demonstrated to be available for use at some specified future date.[19]

The technical support document does not mention some more aggressive options. These include a fuel-neutral standard or a technology-based standard. For example, for carbon dioxide emissions from a newly-constructed power plant, a fuel-neutral standard could follow the example set by the 1997 and 2005 NSPS for nitrogen oxides (NOx) and the 2005 NOx NSPS for modified existing sources. Under those regulations, the NOx emissions standard is the same, regardless of the fuel burned—solid, liquid, or gaseous.[20] This standard is much more expensive for coal-fired facilities to comply with than for natural-gas-fired facilities, thus encouraging the lower-carbon gas-fired technologies. Likewise, EPA could choose to set a newly-constructed power plant standard based on the performance of natural gas burned in a combined-cycle configuration—the fuel and technology of choice for construction of new power plants for the last two decades. If EPA wanted to encourage the rollover of the existing coal-fired power plant fleet to natural gas, nuclear, or renewable sources, it could apply a fuel-neutral standard to modified sources as well. For example, a $CO_2$ emission standard of 0.8 lb. per kilowatt-hour output could be met by a new natural gas-fired, combined-cycle facility, as well as any non-emitting generating technology, such as nuclear power or renewables. In contrast, the standard would require a 60% reduction in emissions from a new coal-fired facility—forcing the development of a carbon control technology, such as carbon capture and storage (CCS), in order for a new coal-fired facility to be built or modified.

The viability of these options, or even more aggressive technology-forcing standards, would depend on how EPA determined whether a technology had been "adequately demonstrated" and its assessment of the seriousness of the technology's costs and energy requirements. As NSPS and BACT determinations are linked, EPA's NSPS determination will have a substantial impact on state-determined BACT actions.

## Clean Air Act Advisory Committee GHG BACT Workgroup

After EPA responded to the Supreme Court's decision in *Massachusetts v. EPA* in April 2009 with a finding that GHGs endangered public health and welfare, its Clean Air Act Advisory Committee (CAAAC) established a Climate Change Work Group in October 2009. The Work Group originally consisted of 35 members, representing a wide variety of industries, state and local government, and environmental and public health organizations. EPA asked the group to:

> discuss and identify the major issues and potential barriers to implementing the PSD Program under the CAA for greenhouse gases. The Work Group should focus initially on the BACT requirements, including information and guidance that would be useful for EPA to provide concerning the technical, economic, and environmental performance characteristics of potential BACT options. In addition the Work Group should identify and discuss approaches to enable state and local permitting authorities to apply the Act's criteria in a consistent, practical and efficient manner.[21]

Over a six-month period, EPA requested two reports from the Group: (1) a relatively brief interim report that identified informational needs of permitting authorities and affected sources; and (2) a final report that contained recommendations to EPA addressing the potential barriers with implementing BACT for GHGs. However, the Work Group had a basic disagreement on the scope of its work. Some felt that the Work Group should assume that the BACT process would apply to GHGs in the same manner that it does for criteria air pollutants, arguing that existing BACT case law and EPA guidance were sufficiently broad to address GHGs. Others argued that the Work Group should explore whether another approach to BACT and GHGs would be more appropriate. The split was resolved by the Work Group agreeing to a two-phase strategy with the first phase focused on providing recommendations to EPA based on current BACT practices and procedures, and a second phase focused on possible alternative or supplementary approaches to GHG BACT.

### *Phase I Report*

Forming four subgroups to address Phase 1 issues, the Work Group found consensus on the following points:[22]

1. *Affected Sources*: BACT should continue to apply to new units and existing units that are undergoing a physical or operational change.
2. *Criteria for Determining Feasible Control Technologies*: Three overarching recommendations for EPA: (a) expand its RACT[23]-BACT-LAER Clearinghouse to include information on GHG related project activities, expand its Office of Research and Development (ORD) GHG mitigation database, and include foreign activities; (b) explore ways to encourage adoption of innovative GHG control technologies; and (c) provide guidance on a sector-by-sector basis regarding evaluating energy efficiency in a BACT analysis.
3. *Criteria for Eliminating Technologies*: With respect to the three basic criteria: (a) *Environmental impacts*: EPA should continue to allow permitting authorities to consider the overall environmental impacts of a GHG application. These impacts include effects on criteria air pollutant levels, water-related impacts, threatened or endangered species, hazardous and solid waste effects, and soils and vegetation. (b) *Energy impacts*: Scope of BACT assessment for energy efficiency is very important. (c) *Economic Impacts*: BACT analysis should be done on a carbondioxide-equivalent basis.
4. *Needs of States and Stakeholders*: Several areas of recommendations were agreed to with respect to (a) timely communications with all stakeholders; (b) EPA guidance on appropriate cost methodologies, approaches, and technologies for GHG reductions; (c) steps to expedite, streamline and provide certainty to the BACT determination process; (d) guidance on netting of GHG emissions; and (e) training for permitting agencies, regulated community and other stakeholders.

Despite many areas of agreement, the Work Group did not provide recommendations with respect to specific control methods or cost thresholds. Carbon Capture and Sequestration (CCS) was discussed by the group, which noted that feasibility would have to be determined on a case-bycase basis. However, there was no attempt to determine the number of CCS systems that must be in use, or whether there must be commercial orders (and how many) before CCS is considered available. The cost thresholds suggested by group members ranged for $3-$ 15 per ton of carbon dioxide equivalence ($CO_2e$) to $30-$150, with others opposed to any fixed monetary values in favor of EPA guidance to permitting authorities on cost-effectiveness values based on the status of various technologies. In addition, no agreement was achieved on an approach in determining the carbon neutrality of biomass, or on the advisability or legality of permits conditioned on the future availability of a control technology or measure.

## *Phase II Report*

At the request of EPA, the Work Group focused its Phase 2 efforts on encouraging development of energy efficient measures and how the Clean Air Act's Innovative Control Technology (ICT) waiver could be used to promote technology development and application.[24] As with the previous report, the expanded 45-member Work Group did not explore specific control methods or cost thresholds.

With respect to better incorporating energy-efficient processes and technologies into the current top-down BACT process, the Work Group provided an analytical framework for determining and evaluating such technologies and processes. The five steps outlined are as follows:

1.  *Identify energy efficient options.* Examples cited include (a) comparing a unit's energy performance with a benchmark to reveal any additional energy efficiency possibilities; (b) identifying options demonstrated overseas, but not in the United States; (c) examining combinations of technologies and/or processes.
2.  *Eliminating technically infeasible options.* Technically infeasible energy efficiency measures should be eliminated in a manner consistent with current EPA guidance. The review of options may include reliability and operational characteristics of the alternative.
3.  *Rank technically feasible energy efficient measures.* Options should be ranked according to specific GHG $CO_2e$ reduction potential for the new facility.
4.  *Evaluate most effective energy efficient measures.* Evaluation should include the energy efficiency and GHG impacts, economic impacts, and environmental impact on other pollutants and media (water, solid waste, etc). Permitting authorities need to assess the tradeoffs between all BACT analyses for each regulated pollutant. The analysis should include a clear justification for elimination of any top candidates.
5.  *Incorporate energy efficient measures into GHG BACT emissions limit.* Besides appropriate monitoring and compliance determination methods, other elements that could be considered in setting the limits include (a) performance standards; (b) operating limits; (c) work practice standards; and (d) design requirements.

With respect to the level at which BACT analysis should occur (individual equipment, entire production line, or entire facility) the Work Group did not reach any specific consensus.

The Work Group's discussion of introducing ways to encourage inherently efficient and lower emitting processes and practices for GHGs focused primarily on the existing Innovative Control Technology (ICT) waiver provided under the Clean Air Act.[25] EPA may grant an ICT waiver from otherwise applicable NSPS requirements to encourage development of an "innovative technological system or systems of continuous emission reduction" that has a substantial likelihood of achieving greater, or at least equivalent, emission reductions than current technology with lower energy, economic, or nonair environmental impacts. The waiver applies to the portion of the source on which it is installed and applies to the deadline for compliance, not the underlying NSPS. The deadline extension is up to seven years after the waiver is granted, or four years after a source commences operation (whichever is earlier). If the technology fails, the waiver may be extended up to three years to allow the facility to comply with the NSPS through other means. EPA may grant only sufficient waivers for a given technology to ascertain whether the technology works and satisfies the energy, economic, and environmental impact criteria.

Although the statutory language applies to NSPS, waiver authority has been included in PSD-BACT regulations since 1980. The addition of BACT to the waiver provisions reflects the interaction between NSPS and BACT. In addition, EPA draft NSR guidance has narrowly defined the need for multiple waivers for an individual technology: "as a practical matter, ... granting of additional waivers to similar sources is highly unlikely since the subsequent applicants are no longer 'innovative.'"[26] The PSD-BACT ICT waiver has been rarely used.

The Workgroup had several recommendations to encourage the use of ICT waivers, consistent with existing statutory authority. In urging EPA to make the ICT waiver more attractive to facilities seeking permits, the Workgroup made the following recommendations:[27]

1. EPA should disavow its current guidance that an ICT waiver is available for only one application of a technology. Multiple waivers for a technology should be allowed as appropriate to encourage commercialization of the technology;

2. EPA should formally and publicly state its views about the waiver's availability in terms of a technology's deployment status and breadth of use among different types of facilities;
3. EPA should reevaluate the maximum time a waiver can be issued for, on a caseby-case basis;
4. EPA should work expeditiously with permitting authorities that wish to issue permits with limits based on innovative technologies (including waivers as needed) and foster information sharing to encourage flexibility in encouraging new and innovative technologies.

## EPA's GHG Guidance

On November 10, 2010, the EPA released its package of guidance and technical information to assist local and state permitting authorities in implementing PSD and Title V permitting for greenhouse gas emissions.[28] The four primary elements of the guidance with respect to PSD permitting are:

1. Basic guidance on implementing PSD permitting, entitled: *PSD and Title V Permitting Guidance for Greenhouse Gases,* that focuses on the five-step permitting process favored by EPA, with particular attention given to determining BACT.
2. GHG Control Measures White Papers that summarize basic technical information on control techniques and measures to reduce GHG emissions from various industries. The information is general and does not define BACT for any sector. The November 10, 2010 release includes white papers for electric generating units, large industrial/commercial/institutional boilers, pulp and paper, cement, nitric acid, and iron and steel.
3. GHG Mitigation Strategies Database that provides data on performance, cost, and environmental effects on available and developing GHG control measures. Data are currently available for electric generating and cement production.
4. Enhancements to the Control Technology Clearinghouse to assimilate information and decisions about GHG control measures required by permits used by state and local authorities.

This package is not a formal rulemaking, although EPA is taking public comment on it and may release a refinement of the package if warranted. EPA will focus on comments pertaining to calculations and other technical errors.

Because the guidance is not a formal rulemaking action, while EPA states it will consider other comments as the process proceeds, it does not intend to formally respond to them.

## What Guidance Did EPA Give?

EPA retains the basic five-step process it has recommended to state and local authorities for 20 years.[29] The primary foci of the EPA guidance package are on state discretion in determining BACT and on energy efficiency as the most likely result of a GHG BACT analysis. These foci are evident through EPA's guidance for each of the five steps. The highlights of that guidance are discussed below.

### *Step 1 Guidance*

The focus of EPA's Step 1 guidance is on inclusion of all potentially applicable control alternatives. In determining inclusion, the guidance discusses the definition of "redefining" (i.e., whether a potential control measure changes the fundamental purpose of the proposed source); the use of energy efficiency measures in BACT analysis; and the possibility for the installation of carbon capture and storage technology under BACT.[30]

A limiting factor in determining the Step 1 list of alternatives is whether an option would fundamentally redefine the nature of the source proposed. If an option is determined by permitting authorities to fundamentally redefine a source, they may exclude it from further consideration. The issue of what alternatives redefine a source has been a subject of contention over the past several years. In three highly publicized cases involving coal-fired power plants— Sithe Global Power's Desert Rock Energy Facility, Deseret Power Electric Cooperative's Bonanza Power Plant, and American Electric Power's John W. Turk Jr. Power Plant—state and regional EPA offices chose not to include integrated gasification combined-cycle technology (IGCC) as a control option in the initial BACT analysis because, based on a 2005 EPA opinion, the technology would redefine the source proposed (which are facilities based on steam boiler technology).[31]

Various petitioners appealed these decisions to EPA's Environmental Appeals Board (EAB). In 2009, the EAB issued a series of opinions that rejected the 2005 EPA opinion and remanded these cases back to their respective state or EPA regional offices for further review. Reiterating its opinion in the Desert Rock case, the EAB stated in the John W Turk Jr. case:

However, as was the case with EPA Region 9's response to comment in the Desert Rock permit, the ADEQ [Arkansas Department of Environmental Quality] has not in fact followed EPA's interpretation on this issue because it has not applied the analytical framework outlined by the EAB in a prior decision, despite citing to that decision as part of its rationale.

ADEQ has thus made the same error as EPA Region 9 by not taking a hard look at how AEP defined its project and to "discern which design elements were inherent to that purpose and which design elements could be changed to achieve pollutant emissions reductions without disrupting [the applicant's] basic business purpose." See Desert Rock, Slip Op. at 69.[32]

In the EPA guidance, EPA suggests permitting authorities look to the EAB decision, stating: "in assessing whether an option would fundamentally redefine a proposed source, EPA recommends that permitting authorities apply the analytical framework recently articulated by the Environmental Appeals Board."[33] Indeed, EPA states clearly that it "no longer subscribes to the reasoning used by the Agency in a 2005 letter to justify excluding IGCC from consideration in all cases on redefining the source grounds."[34]

While EPA recommends the EAB framework, it does not restrict states to it: "The "redefining the source" issue is ultimately a question of degree that is within the discretion of the permitting authority."[35] This discretionary authority is evident as the guidance document discusses "clean fuel" options, such as whether a natural gas electric generating facility is a control option for a proposed coal-fired electric generating facility:

> For example, when an applicant proposes to construct a coal-fired steam electric generating unit, EPA continues to believe that permitting authorities can show in most cases that the option of using natural gas as a primary fuel would fundamentally redefine a coal-fired generating unit. [footnote omitted] Ultimately, however, a permitting authority retains the discretion to conduct a broader BACT analysis and to consider changes in the primary fuel in Step 1 of the analysis.[36]

EPA does believe that energy efficiency measures will have a significant role in GHG BACT analysis. In its guidance, EPA divides energy efficiency measures into two categories. The first category includes technologies and/or processes that maximize the overall efficiency of the proposed facility. In examining these alternatives EPA recommends authorities focus on improving the efficiency of large components and on suites of techniques that can be judged against established benchmarks.

The second category of energy efficiency options includes options that improve the use of thermal energy and electricity that is generated and used on site.

In contrast, EPA classifies CCS as an add-on pollution control technology that is available for all $CO_2$ emitting facilities, such as coal-fired power plants. Therefore, EPA believes it should be listed in Step 1 for such facilities. However, EPA believes it will likely be eliminated in later steps: "Many other case-specific factors, such as the technical feasibility and cost of CCS technology for the specific application, size of the facility, proposed location of the source, and availability and access to transportation and storage opportunities, should be assessed at later steps of a top-down BACT analysis."[37]

### Step 2 Guidance

EPA's specific GHG guidance for Step 2 is primarily focused on CCS.[38] Once again, EPA makes clear that it is the permitting authority that is responsible for deciding technical feasibility:

> Evaluations of technical feasibility should consider all characteristics of a technology option, including its development stage, commercial applications, scope of installations, and performance data. The applicant is responsible for providing evidence that a potential control measure is technically infeasible. However, the permitting authority is responsible for deciding technical feasibility. The permitting authority may require the applicant to address the availability and applicability of a new or emerging technology based on information that becomes available during the consideration of the permit application.[39]

With respect to CCS, EPA expects this step to be a major barrier: "While CCS is a promising technology, EPA does not believe that at this time CCS will be a technically feasible BACT option in certain cases."[40] Particularly, technical hurdles with transport and storage of $CO_2$ may prevent the technology from being reasonably installed and operated in many cases.

### Step 3 Guidance

EPA's specific GHG guidance for Step 3 is primarily focused on energy efficiency measures.[41] Specifically, EPA notes that the concept of overall control effectiveness will need to be refined to ensure that the package of measures with the lowest net emissions is the top-ranked measure. EPA suggests that the ranking be based on the options' net output-based emissions (i.e., include both the emissions reduced and any emissions created by the operation of the control measure to determine net result).

### *Step 4 Guidance*

Step 4 is the critical step for many control options. As stated by EPA:

> In conducting the energy, environmental and economic impacts analysis, permitting authorities have "a great deal of discretion" in deciding the specific form of the BACT analysis and the weight to be given to the particular impacts under consideration. [footnote omitted] EPA and other permitting authorities have most often used this analysis to eliminate more stringent control technologies with significant or unusual effects that are unacceptable in favor of the less stringent technologies with more acceptable collateral environmental effects. However, EPA has also interpreted the BACT requirements to allow for a more stringent technology to remain in consideration as BACT if the collateral environmental benefits of choosing such a technology outweigh the economic or energy costs of that selection. [footnote omitted] ... The same principle applies when assessing technologies for controlling GHGs.[42]

In comparing the environmental impacts of GHG emissions and other regulated pollutants, EPA recommends focusing on the relative levels of emissions rather than on endpoint impacts. Because of the global nature of climate change, the impact of any individual project is likely to be slight. Thus, EPA recommends that the permitting authorities focus on the amount of GHG emissions reductions that would be gained or lost by a specific control measure and how that compares with the collateral increase for other regulated pollutants. EPA states that relatively small collateral increases of another regulated pollutant need not be of concern, "unless even that small increase would be significant, such as a situation where an area is close to exceeding a NAAQS or PSD increment and the additional increase could push the area into nonattainment."[43]

In discussing economic impacts, EPA notes that because of the large volumes of GHGs created by many projects compared with their emissions of other regulated pollutants, it is reasonable that cost effectiveness numbers (in $/ton of $CO_2e$) for GHG control measures will be "significantly lower" than those for other regulated pollutant controls.[44]

As part of its continuing commentary on CCS, EPA states that "even if not eliminated in Step 2 of the BACT analysis, on the basis of the current costs of CCS, we expect that CCS will often be eliminated from consideration in Step 4 of the BACT analysis, even in some cases where underground storage of the captured $CO_2$ near the power plant is feasible."[45] EPA notes that there may be cases where the economics do work out and that future research and development may make CCS more viable in the future.

With respect to energy issues, EPA recommends that permitting authorities consider the impact that a particular control measure would have on the amount of energy that would be produced at an offsite location (e.g., a utility power plant) to support the operation of the proposed facility.

### Step 5 Guidance

Initially, EPA expects many new GHG permits to focus on energy efficiency. In line with that expectation, EPA "encourages permitting authorities to consider establishing an output-based BACT emission limit, or a combination of output- and input-based limits, whenever feasible and appropriate to ensure that BACT is complied with at all levels of operation."[46] EPA notes that in addition to this limit, permits can also include work practice requirements focused on energy efficiency as part of the BACT analysis.

## What about Coal?

The guidance above does not prescribe a specific technology or cost threshold for coal-fired electric generating units. Among the points it does make are the following:

- IGCC does not "redefine" a coal-fired source and should be included in Step 1. States have the discretion with respect to its evaluation under Step 4.
- CCS is an available add-on control measure that should be included in Step 1. EPA expects CCS will generally be eliminated from consideration under Step 2 or Step 4. However, states have the discretion with respect to its evaluation under Step 2 and Step 4.
- Natural gas substitution for coal in a facility is generally considered by EPA to be an option that redefines the source and thus can be excluded under Step 1. However, states have the discretion to determine whether they believe it redefines the source and to evaluate it accordingly.

EPA's focus on energy efficiency for coal-fired facilities is evident in the White Paper that accompanies the guidance document.[47] The White Paper focuses entirely on energy efficiency improvements and carbon capture and storage technologies. Other alternatives—such as co-firing with biomass or natural gas, or natural gas substitution—are not discussed. The White Paper provides an example of a coal-fired unit BACT analysis based on the 2007 air permit application of Consumers Energy Company for an 830 MW supercritical pulverized-coal (PC)-fired unit. During the analysis, the applicant provided an analysis that compared five generating technologies: (1) supercritical PC-fired unit with and without CCS, (2) subcritical PC-fired unit without CCS, (3) subcritical and supercritical circulating fluidized bed (CFB) units without CCS; (4) IGCC unit without CCS, and (5) ultra-supercritical PC-fired unit without CCS. EPA notes that of the units without CCS, the ultra-supercritical PC-fired unit had the lowest projected heat rate and the lowest GHG emissions rate. However, despite stating that ultra-supercritical units "burning various coal ranks are being widely deployed throughout the world" EPA cites a 2007 NETL study to say that "the availability and reliability of materials required to support the elevated temperature environment for high sulfur or chlorine applications, although extensively demonstrated in the laboratory, has not been fully demonstrated commercially."[48] Apparently, the state of Michigan agreed and approved the permit based on the original supercritical design in December 2009.

## What about Innovative Controls?

The new guidance from EPA on innovative technologies is two-fold.[49] First, EPA notes the existence of the innovative control technology waiver for applicants evaluating the use of an innovative technology. Second, EPA reverses its 1990 guidance, and states it will consider granting of multiple waivers for the same or similar technology being proposed at different locations.

## ILLUSTRATING HOW IT MAY WORK: CALIFORNIA'S GHG PERMITTING OF A NATURAL GAS COMBINED-CYCLE POWER PLANT

In February 2010, California's Bay Area Air Quality Management District (BAAQMD) finalized the nation's first PSD permit that includes GHGs in its BACT analysis.[50] The Russell City Energy Center is a 612 MW natural gas-fired combined cycle project to be constructed in Hayward, CA, by an affiliate of Calpine Corporation. In addition to agreeing to greenhouse gas emission limitations, Calpine will also have to install dry low NOx combustors, selective catalytic reduction (SCR), and an oxidation catalyst to limit emissions of other pollutants (particularly NOx). Petitions to the EAB by environmental groups challenging the permit were rejected by the EAB on November 18, 2010.[51]

The permit places limits on greenhouse gas emissions in carbon dioxide equivalence from the facility's two gas turbines and heat recovery steam generators (i.e., the combined-cycle), its fire pump diesel engine, and its five circuit breakers. For the combined-cycle, the limitations are 242 metric tons hourly, 5,802 metric tons daily, and 1,928,182 metric tons annually.[52] In addition, the heat rate of the power plant is not allowed to exceed 7,730 Btu per kilowatt-hour.[53] If the power plant maintains the efficiency required by the permit, it would be allowed to operate at about an 85% capacity factor.

In making this determination, the five-step BACT approach was used. In identifying potential control technology, the technologies identified were thermal efficiency and carbon capture and storage. Some commenters argued in favor of non-fossil electricity generation as an alternative to the proposed plant. However, BAAQMD noted that the 1990 draft EPA guidance does not require consideration of non-fossil-fuel-fired alternatives and deferred to the California Energy Commission. The CCS option was eliminated in step 2 as CCS was considered not commercially available, and no appropriate storage option had been demonstrated.[54] This left efficiency as the only option to achieve GHG reductions.

## HOW MUCH GUIDANCE IS THERE?

For those looking for bright lines and specific recommendations with respect to GHG BACT technologies, particularly with respect to coal-fired

# EPA's BACT Guidance for Greenhouse Gases from Stationary Sources 45

facilities, the released package does not provide them. Indeed, EPA's supplemental "Questions and Answers" release on the guidance seems to stress that it did not draw such conclusions.[55] For example:[56]

- Do these tools identify BACT for specific types of industrial facilities? No.
- Does this guidance say that fuel switching (coal to natural gas) should be selected as BACT for a power plant? No.
- Does this guidance say that carbon capture and storage (CCS) should be selected as BACT? No.

Likewise, the guidance provides no cost thresholds for permitting authorities to consider in determining the economic impacts of alternatives or propose a new approach to selecting BACT for GHG emissions. Instead, the guidance focuses on the discretionary authority that states have in determining BACT; discretion that ensures that BACT will continue to be determined on a case-by-case basis, with states differing in some cases in what they consider appropriate control measures and what constitutes BACT. Whether industry will find such discretion provides sufficient regulatory certainty for it to invest billions in new plant remains to be seen.

In short, the EPA GHG guidance is a simple expansion of the five-step BACT process that has been used for two decades to include greenhouse gases. Whether that is an adequate response will be determined by applicants, state authorities, and future EPA regulatory actions under related parts of the act, such as Section 111 (NSPS), to which BACT is linked.

## End Notes

[1] GHGs addressed by EPA include four different gases and two categories of substances: the individual gases are carbon dioxide ($CO_2$), methane, nitrous oxide (N2O), and sulfur hexafluoride (SF6); the two categories are hydrofluorocarbons (HFCs) and perflourocarbons (PFCs). Each of these substances has a different global warming potential. To facilitate analysis, the emissions of each are converted to the equivalent amount of $CO_2$ emissions, based on how potent the substance is as compared to $CO_2$. For example, SF6 has a global warming potential 22,800 times as great as $CO_2$. SF6 emissions accounted for 16.5 million metric tons of $CO_2$-equivalent in 2007, although actual emissions expressed as SF6, were only 690 metric tons in that year.

[2] For more information, see CRS Report R40506, *Cars, Trucks, and Climate: EPA Regulation of Greenhouse Gases from Mobile Sources*, by James E. McCarthy

[3] These include more stringent regulation of mountaintop removal mining, standards for disposal of coal combustion waste, cooling water intake rules, steam electric utility effluent guidelines, the Maximum Achievable Control Technology (MACT) rule for hazardous air

pollutants such as mercury, the Clean Air Transport Rule for emissions of sulfur dioxide and nitrogen oxides, and a possible New Source Performance Standard (NSPS) for greenhouse gases. For differing views on the potential impact of these proposed actions, see M.J. Bradley & Associates LLC, and Analysis Group, *Ensuring a Clean, Modern Electric Generating Fleet while Maintaining Electric System Reliability* (August 2010), and North American Electric Reliability Corporation, *2010 Special Reliability Scenario Assessment: Resource Adequacy Impacts of Potential U.S. Environmental Regulations* (October 2010).

[4] Section 169(1) lists 28 categories of sources for which the PSD-NSR threshold is to be 100 tons of emissions per year. For all other sources, the PSD-NSR threshold is 250 tons.

[5] Except those pollutants regulated under Sections 112 (hazardous air pollutants) and 211(o).

[6] Currently, there is no NSPS for any greenhouse gas.

[7] U.S. EPA, "Prevention of Significant Deterioration and Title V Greenhouse Gas Tailoring Rule" *Federal Register* (June 3, 2010), p. 31535.

[8] U.S. EPA, "Prevention of Significant Deterioration and Title V Greenhouse Gas Tailoring Rule" *Federal Register* (June 3, 2010), pp. 31514-31608.

[9] Environmental Protection Agency, *New Source Review Workshop Manual: Prevention of Significant Deterioration and Nonattainment Area Permitting* (draft, October 1990).

[10] Categories taken from: Northeast States for Coordinated Air Use Management (NESCAUM), *NESCAUM BACT Guideline* (June 1991), p. 4.

[11] Environmental Protection Agency, *New Source Review Workshop Manual: Prevention of Significant Deterioration and Nonattainment Area Permitting* (draft, October 1990), p. B-14.

[12] LAER refers to Lowest Available Emissions Rate—essentially technology that represents the "best of the best" emissions control. It is required by new facilities locating in an area that is in non-attainment with one of the country's National Ambient Air Quality Standards (NAAQS).

[13] See Environmental Protection Agency, *New Source Review Workshop Manual: Prevention of Significant Deterioration and Nonattainment Area Permitting* (draft, October 1990), pp. B-10-14; and Northeast States for Coordinated Air Use Management (NESCAUM), *NESCAUM BACT Guideline* (June 1991), p. 4.

[14] For background information on the landmark case, see CRS Report RS22665, *The Supreme Court's Climate Change Decision: Massachusetts v. EPA*, by Robert Meltz.

[15] U.S. EPA, "Regulating Greenhouse Gas Emissions Under the Clean Air Act," 73 *Federal Register* 44354, July 30, 2008.

[16] "Regulating Greenhouse Gas Emissions Under the Clean Air Act," 73 *Federal Register* 44356, July 30, 2008.

[17] Ibid.

[18] *U.S. Environmental Protection Agency, Technical Support Document for the Advanced Notice of Proposed Rulemaking for Greenhouse Gases; Stationary Sources, Section VII* (June 5, 2008), final draft.

[19] 73 *Federal Register* 44490, July 30, 2008.

[20] Under Sec. 60.44Da(d)(1), the 1997-2005 NSPS is set at 1.6 lb. per megawatt-hour gross energy output, based on a 20-day rolling average; it is lowered to 1.0 lb. per megawatt-hour gross energy output for power plants commencing construction after February 28, 2005 (Sec. 60.44Da(e)(1). Under Section 60.44Da(e)(3), the 2005 NSPS for modified sources is at either 1.4 lb./MWh on an output basis or 0.15 lb./MMBtu on an input basis. A fuel-neutral standard is also set for reconstructed power plants.

[21] Climate Change Work Group of the Permits, New Source Review and Toxics Subcommittee, Clean Air Act Advisory Committee, *Interim Phase 1 Report* (February 3, 2010), p. 2.

[22] A compilation of the Work Group's reports can be located on EPA's Website at http://www.epa.gov/oar/caaac/ climatechangewg.html.

[23] RACT refers to Reasonably Available Control Technology—essentially emissions control technology that is readily available for retrofit on existing facilities. RACT is required on

# EPA's BACT Guidance for Greenhouse Gases from Stationary Sources 47

existing facilities located in an area that is in non- attainment for a National Ambient Air Quality Standard.

[24] Climate Change Work Group, New Source Review and Toxics Subcommittee,, Clean Air Act Advisory Committee, *Phase II Report of the Climate Change Work Group of the Permits, New Source Review and Toxics Subcommittee, Clean Air Act Advisory Committee* (August 5, 2010), p. 1.

[25] 42 USC 741 1(j)(1)(A). BACT waiver regulations can be found at 40 CFR 52.21(b)(19), 52.21(v), 51. 166(b)(19) and 51. 166(s).

[26] Environmental Protection Agency, *New Source Review Workshop Manual: Prevention of Significant Deterioration and Nonattainment Area Permitting* (draft, October 1990), p. B-13.

[27] Climate Change Work Group, New Source Review and Toxics Subcommittee,, Clean Air Act Advisory Committee, *Phase II Report of the Climate Change Work Group of the Permits, New Source Review and Toxics Subcommittee, Clean Air Act Advisory Committee* (August 5, 2010), pp. 21-22.

[28] EPA's package of GHG permitting guidance and technical resources can be found at its website at http://www.epa.gov/nsr/ghgpermitting.html.

[29] U.S. EPA, "PSD and Title V Permitting Guidance for Greenhouse Gases" (November 2010), p. 18.

[30] U.S. EPA, "PSD and Title V Permitting Guidance for Greenhouse Gases" (November 2010), pp. 25-34.

[31] U.S. EPA, "PSD and Title V Permitting Guidance for Greenhouse Gases" (November 2010). See EPA's discussion of 2005 decision and its reversal on p. 31, footnote 78.

[32] U.S. EPA, "In the Matter of American Electric Power Service Corporation, Southwest Electric Power Company, John W. Turk Plant, Fulton, Arkansas," Permit Number 2123-AOP-RO (signed December 15, 2009), p. 10.

[33] U.S. EPA, "PSD and Title V Permitting Guidance for Greenhouse Gases" (November 2010), p. 27.

[34] U.S. EPA, "PSD and Title V Permitting Guidance for Greenhouse Gases" (November 2010), p. 31, footnote 78.

[35] U.S. EPA, "PSD and Title V Permitting Guidance for Greenhouse Gases" (November 2010), p. 28.

[36] U.S. EPA, "PSD and Title V Permitting Guidance for Greenhouse Gases" (November 2010), p. 29.

[37] U.S. EPA, "PSD and Title V Permitting Guidance for Greenhouse Gases" (November 2010), p. 34.

[38] U.S. EPA, "PSD and Title V Permitting Guidance for Greenhouse Gases" (November 2010), pp. 34-3 8.

[39] U.S. EPA, "PSD and Title V Permitting Guidance for Greenhouse Gases" (November 2010), p. 35.

[40] U.S. EPA, "PSD and Title V Permitting Guidance for Greenhouse Gases" (November 2010), p. 37.

[41] U.S. EPA, "PSD and Title V Permitting Guidance for Greenhouse Gases" (November 2010), pp. 38-39.

[42] U.S. EPA, "PSD and Title V Permitting Guidance for Greenhouse Gases" (November 2010), p. 42.

[43] U.S. EPA, "PSD and Title V Permitting Guidance for Greenhouse Gases" (November 2010), p. 43.

[44] U.S. EPA, "PSD and Title V Permitting Guidance for Greenhouse Gases" (November 2010), p. 44.

[45] U.S. EPA, "PSD and Title V Permitting Guidance for Greenhouse Gases," (November 2010) p. 43.

[46] U.S. EPA, "PSD and Title V Permitting Guidance for Greenhouse Gases," (November 2010) p. 46.

[47] U.S. EPA, "Available and Emerging Technologies for Reducing Greenhouse Gas Emissions Form Coal-Fired Electric Generating Units," (October, 2010).

[48] U.S. EPA, "Available and Emerging Technologies for Reducing Greenhouse Gas Emissions From Coal-Fired Electric Generating Units" (October, 2010), p. 30.

[49] U.S. EPA, "PSD and Title V Permitting Guidance for Greenhouse Gases" (November 2010), pp. 29-30.

[50] Bay Area Air Quality Management District, "Air District Approves Landmark Permit for Hayward Plant" (February 4, 2010), available at http://www.baaqmd.gov/~/media News%20Releases/2010/020410%20Russell%20City.ashx. The permit and related information on the permit is available on the BAAQMD website at http://www.baaqmd.gov/Divisions/Engineering/Public-Notices-on-Permits/2010/020410-15487/Russell-City-Energy-Center.aspx.

[51] U.S. EPA, "Order Denying Review," In re: Russell City Energy Center, LLC, PSD Permit No. 15487, before the Environmental Appeals Board (November 18, 2010), slip opinion.

[52] Broken down by greenhouse gas, the annual limits are (in metric tons): 1,926,399 for $CO_2$, 675 for CH4, and 1,108 for N2O. See Calpine, *GHG BACT Analysis Case Study: Russell City Energy Center* (updated February 3, 2010), p. 10.

[53] Bay Area Air Quality Management District, Prevention of Significant Deterioration Permit Issued Pursuant to the Requirement of 40 CFR 52.21, Permit Application No. 15487 (approved February 4, 2010), p. 17. Heat input refers to higher heating value (HHV).

[54] Calpine, *GHG BACT Analysis Case Study: Russell City Energy Center* (updated February 3, 2010), p. 6.

[55] U.S. EPA, "Clean Air Act Permitting for Greenhouse Gases: Guidance and Technical Information Questions and Answers" (November 10, 2010).

[56] U.S. EPA, "Clean Air Act Permitting for Greenhouse Gases: Guidance and Technical Information Questions and Answers" (November 10, 2010), pp. 1, 4-5.

In: EPA Regulation of Greenhouse Gases ISBN: 978-1-61470-729-5
Editors: Cianni Marino and Nico Costa © 2012 Nova Science Publishers, Inc.

*Chapter 3*

# CARS, TRUCKS, AND CLIMATE: EPA REGULATION OF GREENHOUSE GASES FROM MOBILE SOURCES

## *James E. McCarthy*

### SUMMARY

As Congress and the Administration considered new legislation to reduce the greenhouse gas (GHG) emissions that contribute to climate change over the last year and a half (a process that has now stalled), the Environmental Protection Agency simultaneously began to exercise its existing authority under the Clean Air Act to set standards for GHG emissions. The Administration has made clear that its preference would be for Congress to address the climate issue through new legislation. Nevertheless, it is moving forward on several fronts to define how the Clean Air Act will be used and to promulgate regulations.

On April 1, 2010, EPA used existing authority under Section 202 of the act to set the first national GHG emission standards: the standards will control emissions from new cars and light trucks beginning in model year 2012. The standards will require cars, SUVs, minivans, and other light trucks to meet combined emissions levels that the agency estimates will average 250 grams/mile of carbon dioxide ($CO_2$) in model year 2016, about a 30% reduction in emissions compared to current levels. The standards will be gradually phased in, with the first reduction targets set for model year 2012.

As part of an agreement brokered by the White House, EPA's standards were issued jointly with fuel economy (CAFE) standards developed by the National Highway Traffic Safety Administration, and the state of California agreed to harmonize state-level GHG emission standards, so that the auto industry would have a single national set of standards to meet.

The key to using the Clean Air Act's authority to control greenhouse gases was for the EPA Administrator to find that GHG emissions are air pollutants that endanger public health or welfare. Administrator Jackson promulgated such an endangerment finding in December 2009. With the endangerment finding finalized, the agency can proceed to regulate emissions from motor vehicles of all kinds. Medium- and heavy-duty trucks are next in line: EPA proposed GHG emission standards for them October 25, 2010.

EPA has received 11 petitions asking that it make endangerment findings and proceed to regulate emissions of greenhouse gases. Ten of the 11 petitions address mobile sources: besides motor vehicles, the petitions cover aircraft, ships, nonroad vehicles and engines, locomotives, and fuels, all of which are covered by Title II of the Clean Air Act. In addition to describing the motor vehicle regulations, this report discusses the range of EPA's authority under Title II and provides information regarding other mobile sources that might be regulated under this authority.

Regulation of GHGs from mobile sources will lead the agency to establish controls for *stationary* sources, such as electric power plants, as well.

## INTRODUCTION

Although much of the debate on climate legislation has focused on cap-and-trade[1] and tax options,[2] establishing greenhouse gas controls is not simply a choice between those two alternatives. A third set of options, using the more traditional regulatory approaches of the Clean Air Act (CAA), is available. These regulatory approaches might be modified through new legislation, but unlike a cap-and-trade system or a carbon tax, regulation does not require new congressional action: the ability to limit GHG emissions already exists under various Clean Air Act authorities that Congress has enacted, a point underlined by the Supreme Court in an April 2007 decision, *Massachusetts v. EPA*.[3]

Thus, while Congress and the Administration have discussed new legal authority (for cap-and-trade, carbon tax, and/or targeted emission controls) the Administration, through the Environmental Protection Agency (EPA), has proceeded to exercise existing authority under the Clean Air Act to begin regulation of greenhouse gas emissions.

The agency has taken several steps to regulate GHG emissions. Nevertheless, EPA Administrator Jackson and others in the Administration have made clear that their preference would be for Congress to address the climate issue through new legislation. In an April 2009 press release, for example, the agency stated, "notwithstanding this required regulatory process, both President Obama and Administrator Jackson have repeatedly indicated their preference for comprehensive legislation to address this issue and create the framework for a clean energy economy."[4] Similar statements have been made on several occasions since that time; in the meantime, though, the agency has concluded that current law, including a 2007 Supreme Court decision discussed below, compels the agency to act.

This report focuses on EPA's recent and potential actions regarding regulation of GHG emissions from mobile sources under Title II of the Clean Air Act. We begin with a brief discussion of the petitions and court action that have led to EPA's regulatory decisions.

## *MASSACHUSETTS VS. EPA* AND ITS EFFECTS

Whether EPA could regulate GHGs through existing Clean Air Act authority was under consideration at EPA for more than a decade before the agency took action. In 1998, during the Clinton Administration, EPA General Counsel Jonathan Cannon concluded in a memorandum to the agency's Administrator that greenhouse gases were air pollutants within the Clean Air Act's definition of the term, and therefore could be regulated under the act.[5] Relying on the Cannon memorandum as well as the statute itself, on October 20, 1999, a group of 19 organizations petitioned EPA to regulate greenhouse gas emissions from new motor vehicles under Section 202 of the act.[6] Section 202 gives the EPA Administrator broad authority to set "standards applicable to the emission of any air pollutant from any class or classes of new motor vehicles" if in her judgment they cause or contribute to air pollution which "may reasonably be anticipated to endanger public health or welfare."

Under the Bush Administration, EPA denied the petition August 28, 2003,[7] on the basis of a new General Counsel memorandum the same day, in which it concluded that the CAA does not grant EPA authority to regulate carbon dioxide ($CO_2$) and other GHG emissions based on their climate change impacts.[8] The denial was challenged by Massachusetts, 11 other states, and various other petitioners in a case that ultimately reached the Supreme Court. In an April 2, 2007, decision (*Massachusetts v. EPA*), the Court found by 5-4 that EPA *does* have authority to regulate greenhouse gas emissions, since the emissions are clearly air pollutants under the Clean Air Act's definition of that term.[9] The Court's majority concluded that EPA must, therefore, decide whether emissions of these pollutants from new motor vehicles contribute to air pollution that may reasonably be anticipated to endanger public health or welfare. When it makes such a finding of endangerment, the act requires the agency to establish standards for emissions of the pollutants.[10]

In nearly two years following the Court's decision, the Bush Administration's EPA did not respond to the original petition or make a finding regarding endangerment. Its only formal action following the Court decision was to issue a detailed information request, called an Advance Notice of Proposed Rulemaking (ANPR), on July 30, 2008.[11]

The Obama Administration's EPA, however, made review of the endangerment issue a high priority. On April 17, 2009, it proposed a finding that GHGs do endanger both public health and welfare and that GHGs from new motor vehicles contribute to that endangerment.[12] These findings were finalized in the December 15, 2009 *Federal Register*.[13]

## STANDARDS FOR NEW LIGHT-DUTY MOTOR VEHICLES

The proposed endangerment finding of April 2009 was followed in a matter of weeks by an announcement that the Administration had reached agreement with nine auto manufacturers and California (which had developed its own GHG emission standards for motor vehicles), as well as other interested parties regarding the major outlines of a joint greenhouse gas/fuel economy rulemaking. As announced by the President, May 19, 2009, EPA and the National Highway Traffic Safety Administration (which administers fuel economy standards for cars and trucks) would integrate corporate average fuel economy (CAFE) standards for new cars and light trucks (collectively known as "light-duty motor vehicles") with national greenhouse gas emission standards to be issued by EPA. The objective of the joint standards is to achieve GHG reduction levels similar to those adopted by California, which harmonized its own standards with EPA's as part of the agreement.[14]

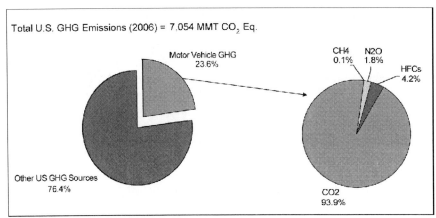

Source: U.S. EPA, March 6, 2009 Draft Deliberative Presentation
Notes: Motor vehicles = passenger cars, light duty trucks, other trucks, buses, and motorcycles, including releases of HFCs from motor vehicle air conditioning. CH4=methane; N2O=nitrous oxide; H FCs=hydrofluorocarbons.

Figure 1. Motor Vehicle Greenhouse Gas Emissions.

Four greenhouse gases are emitted by motor vehicles ($CO_2$, methane, nitrous oxide, and hydrofluorocarbons).[15] According to EPA, emissions of the four gases from motor vehicles (including trucks) accounted for 23.6% of the total inventory of U.S. GHG emissions in 2006. Most of the emissions are in the form of $CO_2$ (see **Figure 1**), which is the product of combusting any fuel containing carbon. Hydrofluorocarbons (HFCs), the chemicals used as coolants in vehicle air conditioning systems, are the second-most important motor vehicle GHG; but, as the figure shows, they are a distant second.

The EPA/NHTSA joint regulations for light-duty motor vehicles were finalized April 1, 2010. They require the vehicles (cars, SUVs, minivans, and other light trucks) to meet combined emissions levels that EPA estimates will average 250 grams/mile of $CO_2$ in model year 2016, about a 30% reduction in emissions compared to current levels. NHTSA has set corresponding fuel economy standards, achieving a combined estimated fuel economy of 34.1 miles per gallon for cars and light trucks by 2016. In both cases, the standards will be gradually phased in, with the first reduction targets set for model year 2012.

In setting the GHG standards, EPA used the concept of a vehicle's "footprint" to set differing standards for different size vehicles. As explained by EPA

These standards are based on $CO_2$ emissions-footprint curves, where each vehicle has a different $CO_2$ emissions compliance target depending on its footprint value (related to the size of the vehicle). Generally, the larger the vehicle footprint, the higher the corresponding vehicle $CO_2$ emissions target. As a result, the burden of compliance is distributed across all vehicles and all manufacturers. Manufacturers are not compelled to build light vehicles of any particular size or type, and each manufacturer will have its own standard which reflects the vehicles it chooses it [sic] produce.[16]

In general, manufacturers are expected to reduce $CO_2$ emissions by improving the vehicles' fuel economy, but they can also take advantage of options to generate $CO_2$-equivalent credits by reducing emissions of hydrofluorocarbons (HFCs) and $CO_2$ through improvements in their air conditioner systems or by the use of idle reduction technologies. Manufacturers will also be allowed to average, bank, and trade emission credits.[17]

The light-duty vehicle rule affects a large group of emission sources that accounts for a significant percentage of total U.S. GHG emissions, but the effectiveness of the standards in reducing total GHG emissions is limited in that they will apply only to new motor vehicles. The auto and light truck fleet turns over slowly: the median survival rate for 1990 cars, for example, was 16.9 years, and that for light trucks was 15.5 years.[18] Given this durability, the impact of GHG standards on the total emissions of the motor vehicle fleet will take a long time to be felt. If historic experience is any guide, reductions in GHG emissions per new vehicle, unless they are very aggressive, may be largely offset by growth in vehicle miles traveled.

## OTHER MOBILE SOURCES

The endangerment finding and emission standards for new light-duty vehicles make it likely that EPA will proceed to set GHG emission standards for other mobile sources of air pollution, especially medium- and heavy-duty trucks. Since the *Massachusetts v. EPA* decision, the agency has received ten petitions asking it to regulate GHGs from other categories, all but one focused on mobile sources and their fuels (see Table 1). These petitions cover aircraft, ocean-going ships and their fuels, motor fuels in general, locomotives, and nonroad vehicles and engines—a category that includes construction equipment, farm equipment, logging equipment, outdoor power equipment, forklifts, marine vessels, recreational vehicles, and lawn and garden equipment.

**Table 1. Petitions for Regulation of Greenhouse Gas Emissions Under the Clean Air Act**

| Date | Subject | CAA Section | Petitioner |
|---|---|---|---|
| 10/20/99 | New Motor Vehicles | 202(a)(1) | International Center for Technology Assessment (ICTA) and 19 other organizations |
| 10/3/07 | Ocean Going Vessels | 213(a)(4) | California Attorney General |
| 10/3/07 | Marine Shipping Vessels and their Fuels | 213(a)(4) and 211 | Oceana, Friends of the Earth, and the Center for Biological Diversity |
| 1/10/08 | New Marine Engines and Vessels | 213(a)(4) | South Coast Air Quality Management District |
| 12/5/07 | Aircraft | 231 | States of California, Connecticut, New Jersey, New Mexico, Pennsylvania, City of New York, District of Columbia, South Coast Air Quality Management District |
| 12/5/07 | Aircraft Engines | 231(a)(2)(A) and 231(a)(3) | Friends of the Earth, Oceana, the Center for Biological Diversity, and the Natural Resources Defense Council |
| 1/29/08 | New Nonroad Vehi-cles and Engines and Rebuilt Heavy-Duty Engines, excluding Aircraft and Vessels | 202(a)(3)(D) and 213(a)(4) | ICTA, Center for Food Safety, and Friends of the Earth |
| 1/29/08 | New Nonroad Vehi-cles and Engines, excluding Aircraft, Locomotives, and Vessels | 202 and 213(a)(4) | States of California, Connecticut, Massachusetts, New Jersey, Oregon, and Pennsylvania |
| 7/29/09 | Fuels Used in Motor Vehicles, Nonroad Vehicles, and Aircraft | 211 and 231 | NYU Law School Institute for Policy Integrity |
| 9/21/09 | Concentrated Animal Feeding Operations | 111(b) and (d) | Humane Society of the United States and 8 other organizations |
| 9/21/10 | Locomotives | 213(a)(5) | Center for Biological Diversity, Friends of the Earth, and ICTA |

Source: U.S. EPA and the petitioning organizations.

The specifics of the Clean Air Act sections that give EPA authority to regulate pollution from these sources vary somewhat, but it is generally believed that the endangerment finding and decision to regulate GHGs in response to the first of the petitions will make it difficult for the agency to avoid regulating at least some of the other categories. With that in mind, we look at other mobile source categories, the authorities provided under Title II for each, and what EPA's use of these authorities for conventional pollutants emitted by these sources indicates with regard to its ability to regulate greenhouse gases.

## Medium and Heavy-Duty Trucks

Section 202(a) of the Clean Air Act, the section that provided authority for the light-duty vehicle GHG standards, requires the Administrator to set "standards applicable to the emission of any air pollutant from *any class or classes of new motor vehicles or new motor vehicle engines*, which in his judgment cause, or contribute to, air pollution which may reasonably be anticipated to endanger public health or welfare" (emphasis added). This authority covers medium- and heavy- duty trucks: in fact, the December 15, 2009, endangerment and cause-or-contribute findings specifically identified the medium- and heavy-duty truck categories as among those that contributed to the GHG emissions for which it found endangerment. As a result, EPA has already begun work on GHG emission standards for these vehicles, and proposed standards on October 25, 2010.

In addition, in the Energy Independence and Security Act of 2007 (EISA, P.L. 110-140), NHTSA was directed to set the first-ever fuel economy standards for medium- and heavy-duty trucks, reflecting the "maximum feasible improvement" in fuel efficiency.[19] Thus, as with light duty vehicles, EPA and NHTSA are cooperating on the setting of standards.

Medium- and heavy-duty trucks are trucks with a gross vehicle weight of 10,000 pounds or more. The largest emitters, tractor-trailers (Class 8 trucks), account for slightly less than 20% of the total number of vehicles, but, because they are heavier and are driven longer distances, they consume 61% of all fuel used by trucks.[20] Presumably, they emit about the same percentage of all trucks' GHGs. Box trucks, which tend to be lighter and are more frequently used in urban settings, are a distant second in terms of GHG emissions. As shown in Table 2, medium- and heavy-duty trucks emitted about 400 million metric tons of GHGs in 2008, about 25% of GHG emissions from motor vehicles. Between 1990 and 2008, emissions from these trucks grew 74%, the fastest growth for any major category of GHG sources. (See Figure 2.)

The EPA Administrator is given substantial leeway in the design and implementation of motor vehicle regulations. The act states that the Administrator may establish categories for purposes of regulation based on "gross vehicle weight, horsepower, type of fuel used or other appropriate factors." In addition, she may delay the effective date of regulations as long as she finds necessary "to permit the development and application of the requisite technology, giving appropriate consideration to the cost of compliance within such period." Using this authority in regulating conventional pollutants, EPA has used weight or power classifications to set differing levels of emission standards, particularly for trucks; it has given manufacturers as much as four years lead time to develop emission controls; and it has set different standards based on the type of fuel an engine uses. Except for specific conventional pollutants mentioned in Section 202, the act does not specify a level of stringency (e.g., best available control technology) for prospective regulations.

Although flexible in many respects, motor vehicle standards have often been used to force the development of new technology. In adopting technology-forcing regulations, EPA has generally followed the lead of California. Because of its more severe air pollution and its pioneering role in establishing motor vehicle emission control requirements in the 1960s, California is allowed to adopt standards more stringent than federal requirements. The state must apply for a waiver of federal preemption under CAA Section 209(b) in order to enforce its more stringent standards, which EPA is to grant if the state meets certain criteria, primarily a showing that the standards are needed to meet "compelling and extraordinary conditions." If California is granted a waiver, other states may adopt identical requirements, thus reinforcing the potential impact of California's technology-forcing standards.

**Table 2. Motor Vehicle GHG Emissions, 2008, by Source Category (million metric tons, $CO_2$-e)**

| Category | Total GHG Emissions | % of Motor Vehicle Total |
|---|---|---|
| Passenger Cars | 632.1 | 39.5% |
| Light Duty Trucks | 552.4 | 34.5% |
| Medium- and Heavy-Duty Trucks | 401.2 | 25.1% |
| Buses | 12.1 | 0.8% |
| Motorcycles | 2.2 | 0.1% |
| Total | 1600.0 | |

Source: U.S. EPA, Inventory of U.S. Greenhouse Gas Emissions and Sinks, 1990-2008, Table 2-15.

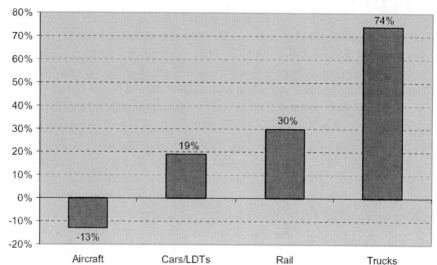

Source: U.S. EPA, Inventory of U.S. Greenhouse Gas Emissions and Sinks, 1990-2008, Table 2-15.

Figure 2. Growth of GHG Emissions from Mobile Sources, 1990-2008.

EPA discussed several potential strategies for reducing GHG emissions from medium- and heavy- duty trucks in its July 2008 Advance Notice of Proposed Rulemaking,[21] including:

1. **Improvements in Engine Technology.** Most trucks, particularly tractor-trailers, are powered by diesel engines, which are already quite efficient; but EPA thinks that a number of small improvements (such as better lubricants and higher cylinder pressure) could increase diesel engine efficiency by up to 20%. For urban trucks, which engage in stop-and-go driving and may idle frequently in traffic, hybrid engine technologies show promise of substantial reductions in emissions.
2. **Eliminating Aerodynamic Drag.** Aerodynamic drag is an important factor in fuel consumption, particularly for tractor-trailers. EPA estimates that drag accounts for 21% of energy consumed by tractor-trailers at 65 miles per hour. The agency has promoted a number of relatively simple redesigns (high roof fairings, side skirts, side fairing gap reducers, aerodynamic mirrors and bumpers) through its SmartWay voluntary program. These measures can have a significant impact on fuel use.
3. **Reducing Rolling Resistance.** Tire rolling resistance accounts for about 13% of energy consumed by tractor trailers, according to EPA. The agency says that 10% or greater reductions in rolling resistance have

already been demonstrated and continued innovation has the potential to achieve larger improvements. In addition to better tires with less rolling resistance, tire inflation indicators can improve fuel efficiency.

4. **Addressing Operational Factors.** Operational factors refer to a wide variety of measures that can reduce truck fuel use, including the installation of speed governors (widely used in Europe and by some fleets in the United States). According to EPA, vehicle speed is the single largest operational factor affecting $CO_2$ emissions from trucks: every one mile per hour increase above 55 mph increases $CO_2$ emissions by more than 1%. Engine idling is another operational factor affecting fuel consumption and GHG emissions. The addition of auxiliary power units or truck stop electrification could eliminate the need for extended idling, reducing emissions.

All in all, EPA stated in its 2008 analysis of the issues, "we see a potential for up to a 40% reduction in GHG emissions from a typical heavy-duty truck in the 2015 timeframe, with greater reductions possible looking beyond 2015...."[22]

EPA and NHTSA jointly proposed GHG and fuel economy standards on October 25, 2010. For a variety of reasons, including agency decisions not to address trailer design issues, the proposed standards would not achieve anywhere near the 40% reduction that the agency found achievable in 2008.

The proposed standards would divide trucks into three main categories: (1) heavy-duty pickup trucks and vans; (2) combination tractors (the power unit of a tractor-trailer combined vehicle); and (3) vocational vehicles.[23]

The standards for heavy-duty pickups and vans use an approach similar to that for light duty vehicles, in which each manufacturer would be required to meet an average standard that would vary depending on its sales mix, with higher capacity vehicles (based on payload, towing capacity, and 4-wheel drive) having less stringent targets. The standards, which would be phased in from 2014 to 2018, are estimated by EPA to cut GHG emissions an average of 17% in diesel vehicles when fully implemented in 2018, and 12% in comparable gasoline-powered vehicles.[24]

For the other categories of trucks, referred to as vocational vehicles or combination tractors, the standards vary significantly depending on the size of the truck. These standards are expected to reduce GHG emissions 7% to 20% for combination tractors and 7% to 10% for vocational vehicles by model year 2017, according to the EPA Regulatory Announcement.[25]

In addition to engine emission standards, the proposal would set a standard for refrigerant leaks, in order to address emissions of HFC greenhouse gases. But trailer design, a major source of efficiency losses (and, thus, higher GHG emissions), is not addressed. According to EPA:

> Trailers are not covered under this proposal, due to the first-ever nature of this proposal and the agencies' limited experience working in a compliance context with the trailer manufacturing industry. However, because trailers do impact the fuel consumption and $CO_2$ emissions from combination tractors, and because of the opportunities for reductions, we are soliciting comments on controlling GHG emissions and fuel consumption from trailers, to prepare a foundation for a possible future rulemaking.[26]

## Ships

Three of the 10 petitions to EPA asking the agency to control greenhouse gas emissions concern ocean-going ships (also referred to as marine engines and vessels) and (in two of the petitions) their fuel. Although there is a wide range of estimates, the International Maritime Organization's consensus is that international shipping emitted 843 million metric tons of carbon dioxide, 2.7% of global $CO_2$ emissions in 2007. Including domestic shipping and fishing vessels larger than 100 gross tonnes, the amount would increase to 1.019 billion metric tons, 3.3% of global emissions.[27]

At these levels, only five countries (the United States, China, Russia, India, and Japan) individually account for a higher percentage of the world total of $CO_2$ emissions.[28]

In addition to the $CO_2$ emissions, the low-quality bunker fuel that ships use and the absence of pollution controls result in significant emissions of black carbon and nitrogen oxides, which also contribute to climate change. Refrigerants used on ships (hydrofluorcarbons and perfluorocarbons—HFCs and PFCs) are also potent greenhouse gases when released to the atmosphere. Thus, the total impact of ships on climate is likely greater than the 3.3% estimate above.

The authority to control pollution from ships is found in Section 21 3(a)(4) of the Clean Air Act, which provides general authority to the Administrator to promulgate standards for emissions other than carbon monoxide, oxides of nitrogen, and volatile organic compounds from "nonroad engines and vehicles."[29] Fuels are regulated separately under Section 211 of the act.

The language of Section 213 is similar to that for new motor vehicles in Section 202, except that in place of the words "cause, or contribute," Section 213 uses the phrase "significantly contribute": if the Administrator determines that emissions of GHGs from ships significantly contribute to air pollution which may reasonably be anticipated to endanger public health or welfare, she may promulgate such regulations as she deems "appropriate." Except for the specific conventional pollutants mentioned in Section 213(a)(2), there is no level of stringency (such as best available control technology) specified for prospective regulations. The Administrator may establish classes or categories of ships for the purposes of regulation. There is no deadline for the promulgation of standards, and in setting them, the Administrator may take into account costs, noise, safety, and energy factors associated with the application of technology.

A wide variety of measures might be undertaken to reduce emissions from shipping, from simple operational measures, such as reducing speed or using cleaner fuels, to various hull and propeller design features that would increase fuel economy. Reducing speed can save substantial amounts of fuel. A.P. Moller-Maersk, which operates the world's largest fleet of containerships, reported that it reduced fuel consumption by its ships 6% in 2008, compared to the fuel used for the same level of business activity in 2007. According to the company, "reducing speed 5-10% does increase the number of days at sea, but reduces both fuel consumption and $CO_2$ emission by more than 1 5%."[30] The petitions also mention improved fleet deployment planning, use of shore-side power while in port, heat recovery systems, the use of sails as supplemental propulsion sources, and NOx controls, such as selective catalytic reduction (SCR) or exhaust gas recirculation, as potential emission control measures.

A complicating factor in the regulation of emissions from ocean-going vessels would be that, for the most part, their GHG emissions occur in international waters, and the sources (the ships) are not registered in the United States: according to California's petition, 95% of the fleet calling on U.S. ports is foreign-flagged. The petitioners assert that these factors are not a bar to EPA regulation, however, citing as precedent a Supreme Court case that held that the Americans with Disabilities Act could be applied to foreign-flagged cruise ships so long as the ADA-required accommodations did not interfere with the ships' internal affairs or require major, permanent modifications to the ships.[31]

In addition to petitioning for regulation of emissions from ships, the petitions from California and from Oceana et al. stated that EPA should regulate the composition of marine shipping vessel fuel to control global-climate-change-related emissions, or should require use of marine diesel fuel oil instead of bunker fuel. The purpose would be to limit the sulfur content of marine fuels and reduce NOx emissions. We discuss EPA's authority to regulate fuels in a separate section below, but note here that EPA, the state of California, and the International Maritime Organization are all moving forward with regulations to limit the sulfur content of bunker fuel for the purpose of reducing conventional pollutants. California's low sulfur fuel requirements went into effect July 1, 2009. In addition, on March 26, 2010, the International Maritime Organization approved an EPA proposal that the entire U.S. coastline except portions of Alaska be designated as an Emission Control Area, subject to lower sulfur limits in bunker fuel.

Sulfur emissions form fine particles of sulfate in the atmosphere, with significant impacts on public health and welfare. Although harmful as a conventional pollutant, sulfur emissions are thought by most experts to be beneficial or at least neutral in the climate context. Sulfates have a cooling effect on the atmosphere, since the particles tend to reflect solar radiation back into space rather than absorbing it. On the other hand, removing sulfur might be necessary to prevent the fouling of pollution control equipment that reduces other pollutants that do lead to warming.

## Other Nonroad Engines

Section 213 can also be used to regulate other nonroad vehicles and engines. A similar endangerment finding would first be required, following which the Administrator may promulgate such regulations as she deems appropriate to control emissions from the classes or categories of nonroad engines that she determines "significantly contribute" to the air pollution that endangers public health or welfare. The Administrator is to take into account costs, noise, safety, and energy factors in setting standards. There is no deadline for setting standards.

The nonroad sector is a broad category that includes construction equipment, farm equipment, forklifts, outdoor power equipment, lawn and garden equipment, and recreational vehicles. This group accounted for 199.7 million metric tons of $CO_2$ emissions in 2007, according to the two petitions requesting regulation (see Table 3), 3.3% of total U.S. emissions of $CO_2$ in that year. According to the ICTA petition, GHG emissions from the nonroad sector increased 49% between 1990 and 2005, a higher rate of emissions increase over the same period than for on-road vehicles (32%), aircraft (3%), boats and ships (36%), and rail (32%).[32]

**Table 3. Nonroad Sector CO$_2$ Emissions, 2007, by Source Category (Million Metric Tons)**

| Category | CO$_2$ Emissions | % of Non road Total |
|---|---|---|
| Construction and Mining Equipment | 63.9 | 32.0% |
| Agricultural Equipment | 39.6 | 19.8% |
| Industrial Equipment | 27.8 | 13.9% |
| Lawn and Garden Equipment | 23.8 | 11.9% |
| Commercial Equipment | 16.4 | 8.2% |
| Pleasure Craft | 15.8 | 7.9% |
| Recreational Equipment | 9.4 | 4.7% |
| Logging Equipment | 1.9 | 1.0% |
| Airport Equipment | 1.0 | 0.5% |
| Railroad Equipment | 0.2 | 0.1% |
| Total | 199.7 | |

Source: ICTA et al., *Petition for Rulemaking Seeking the Regulation of Greenhouse Gas Emissions from Nonroad Vehicles and Engines*. According to the petition, the emissions data were compiled by the Western Environmental Law Center using EPA's nonroad emissions model.

Given their smaller impact on overall emission levels, EPA has been slower to regulate conventional (criteria) pollutants from nonroad engines than from motor vehicles. Many of these engines had few emission control requirements for as many as 25 years after the regulation of automobiles. In the last decade, however, often following the lead of California, EPA has promulgated standards for many nonroad categories. Some of these standards, particularly for diesel-powered equipment and for lawn and garden equipment, have been technology-forcing. Others, such as for snowmobiles, have been less so.

In general, given the wide variety of engine types and sizes and the configurations of the equipment itself, the agency has based its standards on a review of individual subcategories and the technologies available to reduce emissions from specific types of machinery or equipment, rather than applying one across the board standard. Presumably, any GHG standards for this sector would take the same approach.

## Locomotives

On September 21, 2010, EPA received a petition from three environmental organizations to regulate GHG emissions and black carbon from locomotives. In 2008, locomotives emitted 50.6 million metric tons of greenhouse gases. Although this is less than 1% of total U.S. GHG emissions, GHG emissions from railroads increased by 30% between 1990 and 2008, more than twice the rate of increase for total U.S. emissions. In addition, locomotives emit substantial amounts of black carbon (i.e., soot), which is thought to have significant global warming potential through its ability to absorb solar radiation and to reduce the reflectivity of snow and ice. According to a report from NASA's Goddard Institute for Space Studies cited in the locomotive petition, "... black soot may be responsible for 25 percent of observed global warming over the past century."[33] As a result, in addition to requesting that EPA set GHG emission standards for locomotives, the petition asks EPA to set standards for locomotives' black carbon emissions.

The Clean Air Act requires EPA to set emission standards for new locomotives (and new engines used in locomotives) in Section 21 3(a)(5). Unlike almost every other Clean Air Act section dealing with mobile sources, the locomotive subsection does not require an endangerment finding for the Administrator to act. Instead, it requires the Administrator to set standards that achieve the greatest degree of emission reduction achievable through the application of technology which she determines will be available, giving appropriate consideration to cost, noise, energy, and safety factors.

As with the medium- and heavy-duty truck category, EPA discussed several potential strategies for reducing GHG emissions from locomotives in its July 2008 Advance Notice of Proposed Rulemaking (ANPR).[34] The ANPR identified more than 20 strategies for reducing emissions from rail transport, including idle reduction equipment, auxiliary power units, hybrid engines, regenerative braking, and reduction of refrigerant leaks from railcars.

## Aircraft

EPA has also received petitions to regulate GHG emissions from aircraft and aircraft engines. In the United States, aircraft of all kinds are estimated to emit between 2.4% and 3.4% of the nation's total greenhouse gas emissions.[35] When other factors are considered, the impact of U.S. aviation on climate change is perhaps twice that size. These factors include the contribution of

# Cars, Trucks, and Climate    65

aircraft emissions to ozone formation, the water vapor and soot that aircraft emit, and the high altitude location of the bulk of aircraft emissions.

As noted in **Table 1**, two December 2007 petitions requested that EPA address aircraft GHG emissions. Specifically, the petitions asked that EPA make a finding that aircraft GHG emissions endanger public health or welfare, and that the agency adopt regulations that allow a range of compliance approaches: these might include emission limits, operational practices, fees, a capand-trade system, minimizing engine idling time, employing single engine taxiing, or use of ground-side electricity measures to replace the use of fuel-burning auxiliary power units at airport gates.[36]

EPA has authority to regulate emissions from aircraft under Section 231 of the Clean Air Act. The language is similar to that for other mobile sources. It requires the Administrator to issue standards for the emission of any air pollutant from any class or classes of aircraft engines which, in her judgment, causes or contributes to air pollution which may reasonably be anticipated to endanger public health or welfare. The regulations are to take effect "after such period as the Administrator finds necessary ... to permit the development and application of the requisite technology, giving appropriate consideration to the cost of compliance." But compared to other mobile sources, EPA's Clean Air Act authority vis-à-vis aircraft and aircraft engines contains an important difference: the Administrator must consult with the Administrator of the Federal Aviation Administration and the Secretary of Transportation in developing emission standards, and is not allowed to impose new standards if doing so would significantly increase noise and adversely affect safety. The President may also disapprove any such standards if the Secretary of Transportation finds that they would create a hazard to aircraft safety.

Unlike ships, aircraft operating in the United States are generally registered here: EPA has cited data that foreign carriers accounted for only 3% of major carrier operations in the United States in 1999.[37] Thus, whether GHG regulations could be applied to foreign flag carriers might seem to pose less of an issue, at least in terms of whether any potential regulations would address the bulk of the sector's U.S. emissions. On the other hand, international air travel is extremely competitive, and issues of whether regulations can be imposed on foreign carriers have already been raised in the context of the European Union's adoption of cap-and-trade requirements for international aviation. U.S. airlines generally maintain that the imposition of requirements on foreign-flag airlines (i.e., themselves, in the European Union) violates international trade agreements. Their preference is that any controls be

66 James E. McCarthy

negotiated through the International Civil Aviation Organization (ICAO) and be applied equally to all carriers.

EPA has rarely regulated emissions from aircraft without first negotiating international agreements through ICAO. ICAO's regulation of conventional pollutants from aircraft, unlike EPA's regulation of the same pollutants from motor vehicles, has consistently avoided forcing technology. The most recent standards for nitrogen oxides, for example, essentially ratified what the principal aircraft manufacturers had already achieved.[38]

## Fuels

Fuel regulation, whether of bunker fuel, gasoline, or any other type of fuel, is authorized under Section 211 of the Clean Air Act. Section 211 gives the Administrator authority to control or prohibit the manufacture and sale of any fuel or fuel additive if she concludes that its emission products may endanger public health or welfare, or if they will impair to a significant degree the performance of emission control devices. As with the regulation of engines and vehicles, the Administrator is given substantial leeway in the design and implementation of fuel regulations and there is no deadline for their promulgation even after an endangerment finding is made.

GHG emissions from fuels have already been targeted for regulation by the state of California.[39] On April 15, 2010, California's Office of Administrative Law approved regulations to implement the California Low Carbon Fuel Standard, which has been under development since 2007. The standard's goal is to reduce GHG emissions from transportation fuels per unit of energy 10% by 2020. The regulations address emissions from the production, transportation, and consumption of gasoline, diesel fuel, and their alternatives, including biofuels. They envision compliance both through the use of lower carbon fuels and through the development of more efficient, advanced- technology vehicles, such as plug-in hybrids, electric vehicles, and hydrogen fuel cells.

As has been the case with motor vehicles, California has often led the way in the development of cleaner conventional fuels through technology-forcing regulation, with U.S. EPA later adopting similar standards. Thus, many view the Low Carbon Fuel Standard as the prototype of another possible use of existing Clean Air Act authority to regulate greenhouse gas emissions nationally. On July 29, 2009, the Institute for Policy Integrity at NYU Law School petitioned EPA to establish a cap-and-trade system to limit greenhouse gas emissions from fuels used in motor vehicles, nonroad vehicles, and aircraft.

Regulation of fuels would be a way for California or U.S. EPA to obtain reductions from existing vehicles and engines. As noted earlier, the slow turnover of the vehicle fleet means that emission reductions from new vehicles will only gradually affect emission levels from the fleet as a whole. By requiring low carbon fuels, California and EPA could obtain GHG reductions from the entire fleet more quickly.

On the other hand, measuring the carbon content of fuels is more complicated than it may seem, particularly if one considers the life-cycle emissions, including indirect impacts of production. EPA has been embroiled in a controversy over this issue already, as it attempted to develop a methodology for measuring greenhouse gas emissions from biofuels, as required by the Energy Independence and Security Act of 2007 (P.L. 110-140).[40] For regulations implementing this provision, EPA developed and later modified a methodology to measure the GHG effects of indirect land-use changes, such as the switching of land from forest to cropland.[41]

## CONCLUSION

Table 4 summarizes EPA's existing authorities over mobile source GHG emissions and the emissions of each of the sectors discussed in this report. Given the Supreme Court's remand in *Massachusetts v. EPA*, the agency has focused its efforts on motor vehicles (first of all, passenger cars and light trucks), which, as Table 4 shows, account for the majority of mobile source GHG emissions. Having finalized these standards, the agency has now moved on to the development of GHG emission standards for medium- and heavy-duty trucks, the next largest category of emissions. Both of these categories were covered by EPA's December 15, 2009, endangerment finding.

By issuing endangerment findings similar to the one it issued for motor vehicles, EPA could move forward to control GHG emissions from other categories of mobile sources and/or their fuels. On the other hand, once the agency has completed the setting of emission standards for passenger cars, light duty trucks, and medium- and heavy-duty trucks, it will have addressed the categories responsible for more than three-fourths of all mobile source GHG emissions. The next largest category, aircraft, has rarely been the subject of EPA regulation unless the International Civil Aviation Organization

(ICAO) has first agreed on standards. Other mobile source categories are less significant: each accounts for less than 1% of total U.S. emissions.

**Table 4. Categories of Sources whose GHG Emissions Can Be Regulated under Title II of the Clean Air Act (Assuming an Endangerment Finding for the Category)**

| Category | CAA Authority (Section #) | Estimated 2007 GHG Emissions (million tons $CO_2$-e) | % of Total U.S. GHG Emissions |
|---|---|---|---|
| Passenger Cars | 202 | 671.6 | 9.4% |
| Light Duty Trucks | 202 | 569.9 | 8.0% |
| Medium- and Heavy-Duty Trucks | 202 | 425.2 | 5.9% |
| Aircraft (domestic operation) | 231 | 171.8 | 2.4% |
| Construction and Mining Equipment | 213 | 63.9 | 0.9% |
| Ships and Other Boatsa | 213 | 55.2 | 0.8% |
| Locomotives | 213 | 54.3 | 0.8% |
| Agricultural Equipment | 213 | 39.7 | 0.6% |
| Industrial Equipment | 213 | 27.8 | 0.4% |
| Lawn and Garden Equipment | 213 | 23.8 | 0.3% |
| Commercial Equipment | 213 | 16.4 | 0.2% |
| Pleasure Craft | 213 | 15.8 | 0.2% |
| Buses | 202 | 12.5 | 0.2% |
| Recreational Equipment | 213 | 9.4 | 0.1% |
| Motorcycles | 202 | 2.1 | <0.1% |
| Logging Equipment | 213 | 1.9 | <0.1% |
| Airport Equipment | 213 | 1.0 | <0.1% |
| Railroad Equipment | 213 | 0.2 | <0.1% |
| Totala | | 2162.5 | 30.2% |

Source: U.S. EPA and ICTA

a. Does not include international bunker fuel.

Thus, EPA seems more likely to follow up its car and truck standards in two ways: (1) by considering a new round of emission standards for light duty vehicles, and (2) by expanding its focus to stationary sources. On May 21, 2010, President Obama directed the Environmental Protection Agency and the Department of Transportation to develop new fuel economy and greenhouse gas emissions standards for cars and light trucks for model year 2017 and beyond. As has often happened in the development of auto emission standards, EPA will likely be following the lead of California, which expects to adopt GHG emission standards for 2017-2025 model year cars and light trucks by the end of 2010.[42] The California standards would require a waiver of federal preemption from EPA in order to take effect. Emission reductions from these categories might also be addressed through regulation of the carbon content of their fuels.

Meanwhile, EPA can be expected to expand its focus to stationary sources, which account for nearly 70% of the nation's GHG emissions, and within that group to electric power plants. Power plants account for about one-third of all U.S. GHG emissions, a higher percentage of the nation's total than all mobile sources combined. New and modified power plants will automatically be subject to permit requirements and the imposition of Best Available Control Technology as of January 2, 2011, under EPA's interpretation of Section 165 of the Clean Air Act. The agency also appears likely to develop New Source Performance Standards for GHG emissions from electric power plants in the 2011 time period.

## End Notes

[1] For a more detailed discussion of cap-and-trade approaches to GHG emission control, see CRS Report RL33799, *Climate Change: Design Approaches for a Greenhouse Gas Reduction Program*. For additional information on allowance allocation methods, see CRS Report RL34502, *Emission Allowance Allocation in a Cap-and-Trade Program: Options and Considerations*.

[2] For a more detailed discussion of a carbon tax approach to GHG emission control, see CRS Report R40242, *Carbon Tax and Greenhouse Gas Control: Options and Considerations for Congress*.

[3] For a discussion of the Court's decision, see CRS Report RS22665, *The Supreme Court's Climate Change Decision: Massachusetts v. EPA*, by Robert Meltz.

[4] "EPA Finds Greenhouse Gases Pose Threat to Public Health, Welfare/Proposed Finding Comes in Response to 2007 Supreme Court Ruling," Press Release, April 17, 2009, at http://yosemite.epa.gov/opa/admpress.nsf/d0cf6618525a9efb85257359003fb69d/0ef7df 675805295d8525759b00566924.

[5] Memorandum from Jonathan Z. Cannon, EPA General Counsel, to Carol M. Browner, EPA Administrator, "EPA's Authority to Regulate Pollutants Emitted by Electric Power Generation Sources," April 10, 1998.

[6] The lead petitioner was the International Center for Technology Assessment (ICTA). The petition may be found on their website at http://www.icta.org/doc/ghgpet2.pdf.

# 70 James E. McCarthy

[7] The agency argued that it lacked statutory authority to regulate greenhouse gases: Congress "was well aware of the global climate change issue" when it last comprehensively amended the Clean Air Act in 1990, according to the agency, but "it declined to adopt a proposed amendment establishing binding emissions limitations." Massachusetts v. EPA, 549 U.S. 497 (2007).

[8] Memorandum from Robert E. Fabricant, EPA General Counsel, to Marianne L. Horinko, EPA Acting Administrator, "EPA's Authority to Impose Mandatory Controls to Address Global Climate Change Under the Clean Air Act," August 28, 2003.

[9] Massachusetts v. EPA, 549 U.S. 497 (2007). The majority held: "The Clean Air Act's sweeping definition of 'air pollutant' includes '*any* air pollution agent or combination of such agents, including *any* physical, chemical ... substance or matter which is emitted into or otherwise enters the ambient air.... ' ... Carbon dioxide, methane, nitrous oxide, and hydrofluorocarbons are without a doubt 'physical [and] chemical ... substances[s] which [are] emitted into ... the ambient air.' The statute is unambiguous."

[10] For further discussion of the Court's decision, see CRS Report RS22665, *The Supreme Court's Climate Change Decision: Massachusetts v. EPA*, by Robert Meltz.

[11] U.S. EPA, "Regulating Greenhouse Gas Emissions Under the Clean Air Act," 73 *Federal Register* 44354, July 30, 2008. The ANPR occupied 167 pages of the Federal Register. Besides requesting information, it took the unusual approach of presenting statements from the Office of Management and Budget, four Cabinet Departments (Agriculture, Commerce, Transportation, and Energy), the Chairman of the Council on Environmental Quality, the Director of the President's Office of Science and Technology Policy, the Chairman of the Council of Economic Advisers, and the Chief Counsel for Advocacy at the Small Business Administration, each of whom expressed their objections to regulating greenhouse gas emissions under the Clean Air Act. The OMB statement began by noting that, "The issues raised during interagency review are so significant that we have been unable to reach interagency consensus in a timely way, and as a result, this staff draft cannot be considered Administration policy or representative of the views of the Administration." (p. 44356) It went on to state that "... the Clean Air Act is a deeply flawed and unsuitable vehicle for reducing greenhouse gas emissions." The other letters concurred. The ANPR, therefore, was of limited use in reaching a conclusion on the endangerment issue.

[12] U.S. EPA, "Proposed Endangerment and Cause or Contribute Findings for Greenhouse Gases Under Section 202(a) of the Clean Air Act," 74 *Federal Register* 18886, April 24, 2009.

[13] 74 *Federal Register* 66496. Although generally referred to as simply "the endangerment finding," the EPA Administrator actually finalized two separate findings: a finding that six greenhouse gases endanger public health and welfare, and a separate "cause or contribute" finding that the combined emissions of greenhouse gases from new motor vehicles and new motor vehicle engines contribute to the greenhouse gas pollution that endangers public health and welfare.

[14] 74 *Federal Register* 49468, September 28, 2009.

[15] Two other commonly mentioned greenhouse gases, sulfur hexafluoride (SF6) and perfluorocarbons, are not emitted by motor vehicles.

[16] "EPA and NHTSA Finalize Historic National Program to Reduce Greenhouse Gases and Improve Fuel Economy for Cars and Trucks," Fact Sheet, April 2010, p. 3, at http://www.epa.gov/otaq/climate

[17] For additional detail on the EPA/NHTSA joint standards, see CRS Report R40166, *Automobile and Light Truck Fuel Economy: The CAFE Standards*, by Brent D. Yacobucci and Robert Bamberger.

[18] Oak Ridge National Laboratory, for the U.S. Department of Energy, *Transportation Energy Data Book: Edition 27*, 2008, Tables 3.10 and 3.11.

[19] Section 102.

# Cars, Trucks, and Climate 71

[20] National Research Council, *Technologies and Approaches to Reducing the Fuel Consumption of Medium- and Heavy-Duty Vehicles*, Overview Presentation of Report, March/April 2010, p. 26.

[21] U.S. EPA, "Regulating Greenhouse Gas Emissions Under the Clean Air Act," Advance Notice of Proposed Rulemaking, July 30, 2008, 73 *Federal Register* 44453-44458.

[22] Ibid., p. 44454.

[23] In the preamble to the proposed rule, EPA says, "... vocational vehicles consist of a wide variety of vehicle types. Some of the primary applications for vehicles in this segment include delivery, refuse, utility, dump, and cement trucks; transit, shuttle, and school buses; emergency vehicles, motor homes, tow trucks, among others. These vehicles and their engines contribute approximately 15 percent of today's heavy-duty truck sector GHG emissions." EPA and NHTSA, "Greenhouse Gas Emissions Standards and Fuel Efficiency Standards for Medium and Heavy-Duty Engines and Vehicles," pre-publication copy, p. 30, at http://www.epa.gov/otaq/climate

[24] "EPA and NHTSA Propose First-Ever Program to Reduce Greenhouse Gas Emissions and Improve Fuel Efficiency of Medium- and Heavy-Duty Vehicles: Regulatory Announcement," October 2010, p. 5, at http://www.epa.gov/otaq/climate/regulations/420f1 0901 .pdf.

[25] Ibid., pp. 5-7.

[26] Ibid., p. 4.

[27] International Maritime Organization, *Updated Study on Greenhouse Gas Emissions from Ships*, Executive Summary of Phase 1 Report, 1st September 2008, p. 5 at egserver.unfccc.int/seors/attachment/file_storage/6ep77qqvcujba7k.doc. Both estimates exclude emissions from naval vessels.

[28] Oceana, *Shipping Impacts on Climate: A Source with Solutions*, p. 2, at http://www.oceana.org/fileadmin/oceana/uploads/Climate_Change/Oceana_Shipping_Repo rt.pdf.

[29] CO, NOx, and VOCs are regulated under Section 213(a)(3), which requires the imposition of best available control technology, and set a deadline for such regulation.

[30] See *Preparing for the Future*, The A.P. Moller – Maersk Group's Health, Safety, Security and Environment Report 2008, pp. 28-30, at http://media

[31] Spector v. Norwegian Cruiseline, 545 U.S. 119 (2005). In addition, according to the California petition, the United States can and does enforce pollution standards on ships in its territorial waters, "as can be seen by the fact that the National Park Service has imposed air pollutant emissions controls on cruise ships, including foreign-flagged cruise ships (the vast majority of such ships are foreign-flagged), that sail off the coast from Glacier Bay National Park, in Alaska." See People of the State of California Acting by and Through Attorney General Edmund G. Brown, Jr., "Petition for Rule Making Seeking the Regulation of Greenhouse Gas Emissions from Ocean-Going Vessels," October 3, 2007, p. 13. The cited regulations are at 36 CFR 13.65(b)(4). The *Federal Register* citation is 61 *Federal Register* 27008, 27011 (May 30, 1996).

[32] International Center for Technology Assessment, et al., "Petition for Rulemaking Seeking the Regulation of Greenhouse Gas Emissions from Nonroad Vehicles and Engines," January 29, 2008, p. 5.

[33] U.S. National Aeronautics and Space Administration, Goddard Institute for Space Studies, "Black Soot and Snow: A Warmer Combination," December 22, 2003, at http://www.gcrio.org/OnLnDoc/pdf/black_soot.pdf.

[34] U.S. EPA, "Regulating Greenhouse Gas Emissions Under the Clean Air Act," Advance Notice of Proposed Rulemaking, July 30, 2008, 73 *Federal Register* 44463-44464.

[35] The lower percentage includes $CO_2$ emissions from consumption of fuel by military aircraft, general aviation, and domestic operation of commercial aircraft. The higher estimate includes $CO_2$ emissions from international air travel originating in the United States, as well.

[36] For a brief discussion of the petitions, see 73 *Federal Register* 44460, July 30, 2008. Some of these measures, such as minimizing engine idling time, employing single engine taxiing, and use of ground-side electricity measures to replace the use of fuel-burning auxiliary power units, are already widely used by the airlines as fuel-saving measures.

[37] U.S. EPA, Office of Air and Radiation, *Emission Standards and Test Procedures for Aircraft and Aircraft Engines, Summary and Analysis of Comments*, November 2005, p. 10, at http://www.epa.gov/oms/regs/nonroad/aviation/420r05004.pdf

[38] "EPA Proposal to Bring Certain Aircraft Up to International Engine Standard," *Daily Environment Report*, September 30, 2003.

[39] For more information, see http://www.arb.ca.gov/regact/2009/lcfs09/lcfs09.htm. For additional background, see archived CRS Report R40078, *A Low Carbon Fuel Standard: State and Federal Legislation and Regulations*, by Brent D. Yacobucci.

[40] Section 202 of the act mandates the use of "advanced biofuels"—fuels produced from non-corn feedstocks and with 50% lower lifecycle greenhouse gas emissions than petroleum fuel—starting in 2009. Of the 36 billion gallons of renewable fuel required in 2022, at least 21 billion gallons must be advanced biofuel.

[41] For information, see CRS Report R40460, *Calculation of Lifecycle Greenhouse Gas Emissions for the Renewable Fuel Standard (RFS)*, by Brent D. Yacobucci and Kelsi Bracmort.

[42] See "California Seeks to Lead in Strengthening Future Fuel Economy, Emissions Standards," *Daily Environment Report*, April 30, 2010, p. A-6.

In: EPA Regulation of Greenhouse Gases          ISBN: 978-1-61470-729-5
Editors: Cianni Marino and Nico Costa   © 2012 Nova Science Publishers, Inc.

*Chapter 4*

# CLIMATE CHANGE: POTENTIAL REGULATION OF STATIONARY GREENHOUSE GAS SOURCES UNDER THE CLEAN AIR ACT

### *Larry Parker and James E. McCarthy*

## SUMMARY

Although new legislation to address greenhouse gases is a leading priority of the President and many members of Congress, the ability to limit these emissions already exists under Clean Air Act authorities that Congress has enacted – a point underlined by the Supreme Court in an April 2007 decision, *Massachusetts v. EPA*. In response to the Supreme Court decision, EPA has begun the process of using this existing authority, issuing an "endangerment finding" for greenhouse gases (GHGs) December 7, 2009, and proposing GHG regulations for new motor vehicles in the September 28, 2009 *Federal Register*.

On September 30, 2009, the agency took another step toward Clean Air Act regulation of GHGs, proposing what it calls the Greenhouse Gas Tailoring Rule. The rule would define when Clean Air Act permits would be required for GHG emissions from *stationary* sources. The proposed threshold (annual emissions of 25,000 tons of carbon dioxide equivalents) would limit which facilities would be required to obtain permits; for the next six years, the nation's largest GHG emitters, including power plants, refineries, cement production facilities and about two dozen other categories of sources, would be the only sources required to obtain permits. Smaller businesses and almost all farms would be shielded from permitting requirements during this period. By tailoring the permit requirement to the largest sources, EPA says it would focus on about 13,000 facilities accounting for nearly 70% of stationary source GHG emissions.

Like the proposed standards for motor vehicles, the Tailoring Rule is part of a two-track approach to controlling emissions of GHGs. On one track, Congress and the Administration are pursuing new legal authority (for cap-and-trade, carbon tax, or other mechanisms) to limit emissions. At the same time, on a parallel track, the Administration, through EPA, has begun to exercise the Clean Air Act's existing authority to regulate GHGs. Despite EPA's commitment to move forward on this second track, EPA Administrator Jackson and others in the Administration have made clear their preference that Congress address the climate issue through new legislation.

The first step in using the Clean Air Act's existing authority is for the EPA Administrator to find that GHG emissions are air pollutants that endanger public health or welfare. The Administrator proposed this endangerment finding in the April 24, 2009 *Federal Register* and finalized it December 7. With the finding finalized, the agency can (indeed, must) proceed to set GHG emission standards for new motor vehicles, as it proposed to do, September 28.

Motor vehicle GHG standards will lead EPA and state permitting authorities to require permits for stationary sources: language in the Act triggers permitting under the Prevention of Significant Deterioration (P SD) program and Title V of the Act whenever a pollutant is "subject to regulation" under any of the Act's authorities. It is this trigger that the Tailoring Rule addresses.

This report reviews the various options that EPA could exercise to control GHG emissions from stationary sources under the Act. The PSD and Title V permitting requirements that are automatically triggered may be the most immediate point of interest, but an endangerment finding for GHGs would present the agency with other options, as well. Five of these options are discussed in this report. Among these, particular attention should be paid to Section 111 of the Act, which provides authority to set New Source Performance Standards and, under Section 111(d), requires the states to control emissions from existing sources of the same pollutants. As EPA moves forward, Section 111 appears to be the most likely authority it will use to establish emission standards for stationary sources.

# INTRODUCTION

This report was originally published in May 2009, and the majority of the text reflects the authors' analysis of EPA's potential regulation of stationary sources under the Clean Air Act at that time. That analysis has not changed substantially, but since that time, EPA has given several further indications of its intentions with regard to the regulation of stationary sources of greenhouse gases, in congressional testimony, proposed regulations, and proposed guidance. The agency has formally proposed greenhouse gas emission standards for new motor vehicles,[1] and stated that it intends to promulgate such standards by March 31, 2010. An "endangerment finding,," which is a prerequisite for the motor vehicle and other greenhouse gas standards, was finalized December 7, 2009.[2] The agency has stated in several venues that promulgation of motor vehicle GHG standards would make GHGs "subject to regulation" for the purposes of triggering permitting requirements for new and modified stationary sources under the Prevention of Significant Deterioration requirements of Section 165 of the Clean Air Act, and also for the purposes of the operating permit requirements of Title V.[3] And it has proposed a Greenhouse Gas Tailoring Rule to limit the applicability of the PSD and Title V permitting requirements to sources that emit more than 25,000 tons per year of carbon dioxide equivalents.[4]

New legislation to address greenhouse gases is a leading priority of the President and many members of Congress, but the ability to limit these emissions already exists under various Clean Air Act (CAA) authorities that Congress has enacted, a point underlined by the Supreme Court in an April 2007 decision (discussed below). Indeed, the U.S. Environmental Protection Agency (EPA) has already begun the process that could lead to greenhouse gas regulations for new mobile sources in response to court decisions.

When EPA finalizes the regulation of greenhouse gases from new mobile sources, legal and policy drivers will be activated that will lead to regulation of stationary sources as well. The legal drivers are beyond the scope of this report, which is focused on the policy options and control alternatives available to EPA as it uses existing authorities to regulate greenhouse gases from stationary sources.

Stationary sources are the major sources of the country's greenhouse gas emissions. Overall, 72% of U.S. emissions of greenhouse gas come from stationary sources (the remainder come from mobile sources). As indicated in Table 1, relatively large sources of fossil-fuel combustion and other industrial processes are responsible for about one-half the country's total emissions. If

EPA were to embark on a serious effort to reduce greenhouse gas emissions, stationary sources, and in particular large stationary sources, would have to be included. This concentration of greenhouse gas emissions is even more important from a policy standpoint: reductions in greenhouse gas emissions from these sectors are likely to be more timely and cost-effective than attempts to reduce emissions from the transport sector.

This report discusses three major paths and two alternate paths of statutory authorities that have been identified by EPA and others as possible avenues the agency might take in addressing greenhouse gas emissions under existing CAA provisions. After discussing the approaches, we identify categories of control options EPA could consider, including an EPA-coordinated cap-and-trade program. Then we discuss the administrative difficulties in using the Clean Air Act for greenhouse gas control, particularly New Source Review – Prevention of Significant Deterioration and Title V permitting requirements. Finally, we conclude by putting the issue into the context of previous environmental challenges the CAA has faced.

### Table 1. Selected U.S. Stationary Sources of Greenhouse Gases

| Source | 2007 Emissions | % of Total GHGs |
|---|---|---|
| Coal-fired | 1977.7 | 27.8% |
| Natural gas-fired | 374.1 | 5.3% |
| Fuel Oil-fired | 55.4 | 0.8% |
| **Industrial fossil-fuel combustion ($CO_2$, $CH_4$, $N_2O$)** | | |
| Mostly Petroleum refineries, chemicals, primary metals, paper, food, and nonmetallic mineral products | | |
| Coal-fired | 108.1 | 1.5% |
| Natural gas-fired | 385.6 | 5.4% |
| Fuel Oil-fired | 353.3 | 5.0% |
| **Industrial Processes** | | |
| Iron and Steel Production ($CO_2$, $CH_4$) | 74.3 | 1.0% |
| Cement Production ($CO_2$) | 44.5 | 0.6% |
| Nitric Acid Production ($N_2O$) | 21.7 | 0.3% |
| Substitution of Ozone Depleting | 108.3 | 1.5% |
| Substances (HFCs) | | |
| **Other** | | |
| Natural Gas Systems ($CO_2$, $CH_4$) | 133.4 | 1.9% |
| Waste Incineration ($CO_2$, $N_2O$) | 21.2 | 0.3% |
| **TOTAL** | **3657.6** | **51.3%** |

Source: EPA, inventory of U.S. Greenhouse Gas Emissions and Sinks: 1990-2007, April 2009.

# The Entry Point: Massachusetts vs. EPA

A regulatory approach using existing Clean Air Act authorities has been under consideration at EPA for more than a decade. In 1998, EPA's General Counsel, Jonathan Cannon, concluded in a memorandum to the EPA Administrator that greenhouse gases were air pollutants within the Clean Air Act's definition of the term, and therefore could be regulated under the Act.[5] Relying on the Cannon memorandum as well as the statute itself, on October 20, 1999, a group of 19 organizations petitioned EPA to regulate greenhouse gas emissions from new motor vehicles under Section 202 of the Act.[6] Section 202 gives the EPA Administrator broad authority to set "standards applicable to the emission of any air pollutant from any class or classes of new motor vehicles" if in her judgment they contribute to air pollution which "may reasonably be anticipated to endanger public health or welfare."

EPA denied the petition in 2003[7] on the basis of a new General Counsel memorandum issued the same day in which the General Counsel concluded that the CAA does not grant EPA authority to regulate $CO_2$ and other GHG emissions based on their climate change impacts.[8] The denial was challenged by Massachusetts, eleven other states, and various other petitioners in a case that ultimately reached the Supreme Court. In an April 2, 2007 decision (*Massachusetts v. EPA*), the Court found by 5-4 that EPA *does* have authority to regulate greenhouse gas emissions, since the emissions are clearly "air pollutants" under the Clean Air Act's definition of that term.[9] The Court's majority concluded that EPA must, therefore, decide whether emissions of these pollutants from new motor vehicles contribute to air pollution that may reasonably be anticipated to endanger public health or welfare. If it makes this finding of endangerment, the Act requires the agency to establish standards for emissions of the pollutants.[10]

## The Advance Notice of Proposed Rulemaking (ANPR)

For nearly two years following the Court's decision, the Bush Administration's EPA did not respond to the original petition nor make a finding regarding endangerment. Its only formal action following the Court decision was to issue a detailed information request, called an Advance Notice of Proposed Rulemaking (ANPR), on July 30, 2008.[11]

The ANPR occupied 167 pages of the *Federal Register*. Besides requesting information, it took the unusual approach of presenting statements from the Office of Management and Budget, four Cabinet Departments (Agriculture, Commerce, Transportation, and Energy), the Chairman of the Council on Environmental Quality, the Director of the President's Office of Science and Technology Policy, the Chairman of the Council of Economic Advisers, and the Chief Counsel for Advocacy at the Small Business Administration, each of whom expressed their objections to regulating greenhouse gas emissions under the Clean Air Act. The OMB statement began by noting that, "The issues raised during interagency review are so significant that we have been unable to reach interagency consensus in a timely way, and as a result, this staff draft cannot be considered Administration policy or representative of the views of the Administration."[12] It went on to state that "... the Clean Air Act is a deeply flawed and unsuitable vehicle for reducing greenhouse gas emissions."[13] The other letters concurred. The ANPR, therefore, was of limited use in reaching a conclusion on the endangerment issue and, in any event, it presents the views of an Administration no longer in office.

The current Administration made review of the endangerment issue a high priority. On April 17, 2009, EPA proposed a finding that GHGs do endanger both public health and welfare and that GHGs from new motor vehicles contribute to that endangerment.[14] Publication of the proposal in the *Federal Register* on April 24 began a 60-day public comment period. In addition, public hearings were held May 18 in Arlington, VA, and May 21 in Seattle, WA. The endangerment finding was finalized December 7, 2009.

## Potential Implications for Stationary Sources

While there has been considerable speculation in the literature about the meaning of *Massachusetts v. EPA* for stationary sources, there have also been several attempts to invoke the various authorities of the Clean Air Act to begin controlling greenhouse gas emissions from stationary sources.[15] Among the legal initiatives currently underway are the following:

- In 2006, the EPA revised the New Source Performance Standard (NSPS) for electric utilities and other steam generating units without including any $CO_2$ standard, or other requirement. Led by New York, several states filed a petition for review of the new NSPS, challenging the omission of any $CO_2$ requirement. In September 2007 the D.C. Circuit Court of

# Climate Change

Appeals remanded the case back to EPA for further proceedings "in light of *Massachusetts v. EPA*."[16]

- In 2007, EPA Region 8 granted a Prevention of Significant Deterioration (P SD) permit authorizing construction of a waste-coal-fired electric generating plant near Bonanza, Utah. Appealing the decision, the Sierra Club argued to the Agency's Environmental Appeals Board (EAB) that because the Court had found in *Massachusetts v. EPA* that $CO_2$ was an air pollutant under the Act, and that EPA has imposed $CO_2$ monitoring and reporting requirements, the Bonanza plant was required to install Best Available Control Technology (BACT) for $CO_2$ emissions. The EAB rejected the Sierra Club's interpretation of the PSD-NSR language, but remanded it back to Region 8 for reconsideration of a $CO_2$ BACT requirement.[17] In a second case, on February 18, 2009, the EAB remanded a permit issued by the Michigan Department of Environmental Quality for reconsideration of its decision not to regulate $CO_2$ from a new cogeneration boiler at Northern Michigan University.[18] In a third PSD-NSR (New Source Review) case, EPA Region 9 filed a motion with the EAB in April 2009 for a voluntary remand of the PSD permit for the Desert Rock coal-fired power plant in New Mexico to allow for a reconsideration of its permit to include a $CO_2$ limitation. Region 9 wants to reconsider its decision not to require Desert Rock to install "carbon-ready" integrated gasification combined-cycle technology instead of allowing current pulverized-coal technology.[19] The EAB remanded the permit September 24, 2009.
- In 2009, the Environmental Integrity Project, an environmental group, filed a complaint with the D.C. Circuit Court to force the EPA to review nitrous oxide (N2O) emissions from nitric acid plants.[20] The group argues that EPA has not reviewed the NSPS for such plants since 1984, despite the statutory requirements for periodic reviews.

It should be noted that amidst this legal activity and EPA's commitment to move forward with an endangerment finding, EPA Administrator Jackson and others in the Administration have made clear that their preference would be for Congress to address the climate issue through new legislation. In the press release announcing the proposed endangerment finding, the agency stated, "Notwithstanding this required regulatory process, both President Obama and Administrator Jackson have repeatedly indicated their preference for comprehensive legislation to address this issue and create the framework for a clean energy economy." Similar language was used in the press release accompanying the finalization of the endangerment finding, December 7.

# POTENTIAL PATHS FOR GHG STATIONARY SOURCE CONTROL

When looking at the CAA from the point of view of reducing GHGs from stationary sources, three existing paths are available. As indicated in Table 2, the three paths are (1) to regulate GHGs as criteria air pollutants, (2) to regulate GHGs as hazardous air pollutants, or (3) to regulate GHGs as designated air pollutants. Each of these paths are discussed below, along with two lesser explored trails: Section 115 and Title VI.

**Table 2. Simplified Requirements under Title I for Most Stationary Sources**

| | Section 109 (NAAQS) | Section 112 (Air Toxics) | Sections 111(d)/1 29 (Designated Pollutants) |
|---|---|---|---|
| Minimum Controls | **New/Modified Source:** EPA-determined NSPS under Sec. 111 | **New Source:** EPA-determined MACT under Sec. 112(d) | **New/Modified Source:** EPA-determined NSPS under Sec. 111 |
| | **Existing Source:** Depends on area's attainment status/ visibility provisions | **Existing Source:** Less stringent EPA-determined MACT | **Existing Source:** State determination under EPA standards issued under Sec. 111(d) |
| Implementing Provisions | State Implementation Plans under Sec. 110 New Source Review (NSPS, PSD, nonattainment) Sec. 126 Petitions | Statutory list under Sec. 112(b)(1) EPA determination under Sec. 11 2(b)(2) or (b)(3) | Designated Pollutant Plans under Sec. 111(d)/ 129 New Source Review (PSD) |

Notes: NAAQS stands for National Ambient Air Quality Standard and is discussed below. MACT stands for Maximum Achievable Control Technology and is discussed after the discussion of NAAQS.

Climate Change 81

## Path 1. Regulating GHG through National Ambient Air Quality Standards (NAAQS)

### *Importance of NAAQS*

The backbone of the Clean Air Act is the creation of National Ambient Air Quality Standards (NAAQS). The need to attain NAAQS, which are set at levels designed to protect public health without consideration of costs or economic impact, is the driving force behind much of clean air regulation.

The authority for NAAQS is found in Sections 108 and 109 of the Act. Under Section 108, EPA is to identify air pollutants that, in the Administrator's judgment, endanger public health or welfare, and whose presence in ambient air results from numerous or diverse sources. Under Section 109, EPA is required to set NAAQS for the identified pollutants.

Section 109 requires the EPA Administrator to set both primary and secondary NAAQS. Primary NAAQS must be set at a level that will protect public health with an adequate margin of safety. Secondary NAAQS are required to protect public welfare from "any known or anticipated adverse effects associated with the presence of such air pollutant in the ambient air." Public welfare covers damage to crops, vegetation, soils, wildlife, water, property, building materials, etc., and such broader variables as visibility, climate, economic values, and personal comfort and well-being.

Over the years, EPA has identified six air pollutants or categories of air pollutants for NAAQS: sulfur dioxide ($SO_2$), particulate matter (PM2.5 and PM10), nitrogen dioxide ($NO_2$), carbon monoxide (CO), ozone, and lead. These six are referred to as "criteria" pollutants. Each of the criteria pollutants was identified for NAAQS regulation in the 1970s. Since that time, although the specific standards (the allowed concentrations) have been reviewed and modified, no new criteria pollutants have been identified.

### *NAAQS and Controlling GHGs*

If carbon dioxide ($CO_2$) or other greenhouse gases were identified as criteria pollutants, NAAQS would then have to be set. $CO_2$, the most important greenhouse gas, is an air pollutant that EPA has determined endangers both public health and welfare, and its presence in ambient air results from numerous or diverse sources. Thus, it meets the basic criteria of Section 108. But setting a NAAQS for $CO_2$ raises a number of potential issues, four of which are discussed in the following sections.

## Setting a Standard

An initial difficulty would arise in choosing a level at which to set a NAAQS. Primary and secondary NAAQS are expressed as concentrations of the pollutant in ambient air that endanger public health or welfare. For the six current criteria pollutants, the focus has been on setting *primary* (health-based) standards—i.e., identifying a concentration in ambient air above which ambient concentrations of the pollutant contribute to illness or death. These standards are based on both concentration-response studies undertaken in laboratory conditions (often animal studies, but some involving humans), and on epidemiology that demonstrates a correlation between greater exposure to the pollutant and higher rates of morbidity and mortality.

For $CO_2$ at current and projected levels, there are not the same direct linkages between higher concentrations and health as there are for each of the current NAAQS. A person exposed to current ambient levels of $CO_2$ will not be sickened. Nor is it likely that one could demonstrate a connection between $CO_2$ and morbidity or mortality through epidemiology, in part because $CO_2$ concentrations are relatively uniform across the globe and change very slowly. The argument that can be made is more indirect: that higher levels of $CO_2$ are likely over time to cause higher temperatures, and higher temperatures and associated changes in climate-related processes are likely to have health consequences.

If EPA concluded that this connection between $CO_2$, higher temperatures, and human health were sufficient to justify establishing a primary NAAQS, it would still be difficult to pick out a specific $CO_2$ concentration for a standard. Among scientists concerned about greenhouse gas concentrations, some argue for a level of 350 parts per million (ppm) as the concentration that must be attained,[21] others argue for 450 ppm, and some for levels of 550-600 ppm. Current concentrations in the Earth's atmosphere are about 385 ppm, increasing by 1 or 2 ppm per year. The mechanics of implementing a standard will be discussed in greater detail below, but it is important to note here that unless one chose a standard at or below the current ambient level, establishing a primary NAAQS would have no consequence. It is only if ambient concentrations of the pollutant exceed the standard that action must be taken.

A further point regarding the setting of a NAAQS is the importance of distinguishing primary from secondary standards. If one were to set a NAAQS for $CO_2$ or other GHGs, it is perhaps the secondary NAAQS that is most relevant to the discussion. As noted above, secondary NAAQS are designed to prevent damage to crops, vegetation, soils, wildlife, water, property, building

# Climate Change 83

materials, etc. and such broader variables as visibility, climate, economic values, personal comfort and well-being.

EPA—under both Democratic and Republican Presidents—has generally given short shrift to the setting of secondary NAAQS: most have been set at a level identical to the primary standard, with little discussion of the agency's reasoning. In part, this is because secondary NAAQS have no deadlines attached to their attainment and there is no enforcement mechanism or penalty for failure to attain them.

Thus, it would hardly be worth the effort to establish a NAAQS for GHGs unless one could establish a defensible case for a specific primary standard that was below ambient levels. Primary NAAQS, unlike their secondary kin, do have deadlines: there are consequences for a failure to attain them in a timely manner.

## Identifying Nonattainment Areas

If a $CO_2$ or GHG NAAQS were set by EPA, the next step would be to identify nonattainment areas (i.e., areas where ambient concentrations of $CO_2$ and/or other GHGs exceed the NAAQS). The procedure for doing so is specified under Section 107 of the Act. For the six current criteria pollutants, there are distinct local and regional concentrations of each pollutant that can generally be linked to stationary or mobile sources in the area. In some cases, the sources may be relatively distant, with pollutants (or precursors) emitted hundreds of miles away. But with all of the current criteria pollutants, there are significant variations in local and regional concentrations, and only those areas with pollutant readings higher than the NAAQS are designated "nonattainment."

For $CO_2$, this would not be the case. Concentrations are relatively homogeneous across the entire country—indeed, across the world. Thus, the entire United States would need to be designated nonattainment if concentrations exceeded the standard.

## Developing State Implementation Plans

A third element of NAAQS that appears ill-suited to the regulation of GHGs is the mechanism used to bring about compliance with NAAQS, the State Implementation Plan (SIP) provisions in Section 110 and Sections 171-1 79B. SIPs describe the sources of pollution in a nonattainment area and the methods that will be used by the area to reduce emissions sufficiently to attain the standard. They are required to be developed and submitted to EPA for each nonattainment area within three years of its designation.

SIPs build on some national standards (for new motor vehicles and new or modified power plants, for example), but they assume that most sources of the pollution to be controlled are local, and therefore, that the measures needed to reach attainment are measures tailored to local conditions. To the extent that significant emission sources are located in other states, downwind states are authorized under Section 126 to petition EPA for controls on such upwind sources.

If pollution is uniform throughout the country, there is no reason why the measures taken to reduce it should vary from locality to locality. Nor will a nonattainment area be able to demonstrate that its pollution control measures will have any measurable impact on the ambient concentration of most greenhouse gases. Thus, State Implementation Plans tailored to each nonattainment area would be ill-suited to the nature of the problem.

**Attaining the Standard**

It is also unlikely that any state or nonattainment area on its own could demonstrate reasonable further progress toward attainment of the standard (as is required by Section 172), particularly within the 5- to10-year period specified in Section 172 for attainment of a NAAQS. Greenhouse gases accumulate in the atmosphere, and some can take hundreds of years to diminish, even if current global emissions decline. Global emissions are increasing. Individual states and nonattainment areas would have little chance of reversing this trend through any set of actions they might undertake on their own.

Despite all of these difficulties, two groups (the Center for Biological Diversity and *350.org*) petitioned EPA on December 2, 2009, to designate carbon dioxide a criteria air pollutant and set a NAAQS for it at no greater than 350 ppm. They further requested that EPA designate six other greenhouse gases as criteria pollutants and establish pollution caps for them "at science-based levels."[22]

# Path 2. Regulating GHGs through Section 112 as Hazardous Air Pollutants

## *Importance of Section 112*

As revised by the 1990 CAA amendments, Section 112 contains four major provisions: Maximum Achievable Control Technology (MACT) requirements for major sources; health-based standards to be imposed for the

Climate Change 85

residual risks remaining after imposition of MACT standards; standards for stationary "area sources" (small, but numerous sources, such as gas stations or dry cleaners, that collectively emit significant quantities of hazardous pollutants); and requirements for the prevention of catastrophic releases. The MACT and area source provisions would appear to be the most relevant, if GHGs were to be controlled under this section.

The MACT provisions require EPA to set standards for sources of the listed pollutants that achieve "the maximum degree of reduction in emissions" taking into account cost and other nonair-quality factors. MACT standards for new sources "shall not be less stringent than the most stringent emissions level that is achieved in practice by the best controlled similar source." The standards for existing sources may be less stringent than those for new sources, but generally must be no less stringent than the average emission limitations achieved by the best performing 12% of existing sources. Existing sources are given three years following promulgation of standards to achieve compliance, with a possible one-year extension; additional extensions may be available for special circumstances or for certain categories of sources.

In addition to the technology-based standards for major sources of hazardous air pollution, Section 112 requires EPA to establish standards for stationary "area sources" (small, but numerous, sources such as gas stations or dry cleaners, that collectively emit significant quantities of hazardous air pollutants). In setting these standards, EPA can impose less stringent "generally available" control technologies, rather than MACT.

### Section 112 and Controlling GHGs

Could EPA regulate GHG emissions as hazardous air pollutants under Section 112? In its comments on the ANPR, the Bush Administration's Department of Energy stated that "... it is widely acknowledged that a positive endangerment finding could lead to ... the listing of one or more greenhouse gases as hazardous air pollutants (HAP) under section 1 12."[23] EPA, on the other hand, was more circumspect in its analysis, stating:

> The effects and findings described in section 112 are different from other sections of the CAA addressing endangerment of public health discussed in previous sections of today's notice. Given the nature of the effects identified in section 112(b)(2), we request comment on whether the health and environmental effects attributable to GHG fall within the scope of this section.[24]

The language of Section 112 refers to pollutants that may present a threat of adverse human health effects or adverse environmental effects. This language might be broad enough that GHGs could be categorized as hazardous air pollutants and subjected to the regulatory tools provided by the section, but because the section was written to apply to carcinogenic and other toxic air pollutants present in emissions in small quantities, there would be questions as to whether Congress intended the use of the section's authority for pollutants such as GHGs. The legislative history of the Act makes clear that it was designed primarily to regulate pollutants commonly referred to as "air toxics." Hazardous air pollutants are defined as "any pollutant listed pursuant to subsection [112](b)." Congress provided an initial list of 189 hazardous air pollutants in that subsection, and it established criteria and procedures for revising the list in Section 112(b)(2). In the 18 years since the criteria were established, EPA has not added any substances to the list.

The procedures for revising the list provide that the Administrator may do so "by rule," adding pollutants that may present, through inhalation or other routes of exposure, a threat of adverse human health effects, or, through a variety of routes of exposure, adverse environmental effects. The human health effects language is qualified with wording that suggests the type of pollutants Congress had in mind when it drafted this section: substances that include, but are not limited to, ones known or reasonably anticipated to be carcinogenic, mutagenic, teratogenic, neurotoxic, acutely or chronically toxic, or which cause reproductive dysfunction.

The section is also not well-suited to the most common GHGs, such as $CO_2$, that are emitted in very large quantities. For example, it defines a major source as one that emits 10 tons per year or more of any hazardous air pollutant. Annual $CO_2$ emissions in the United States are about 6 billion metric tons, and hundreds of thousands, perhaps millions of sources (including large residential structures) might qualify as major sources if $CO_2$ were listed as a hazardous air pollutant under this section.

Section 112 might be useful, if at all, for regulating small volume chemicals that are very potent greenhouse gases: sulfur hexafluoride (SF6), for example. SF6 has a global warming potential 22,800 times as great as $CO_2$ and accounted for about one-quarter of one percent of total U.S. GHG emissions in 2007, when measured by its global warming potential. SF6 emissions were 16.5 million metric tons of $CO_2$-equivalent in that year. Actual emissions expressed as SF6, however, were only 690 metric tons. Nitrogen trifluoride (NF3), another chemical with low emission levels but high global warming potential, might be another candidate, if EPA chose this regulatory route.

Section 112 generally considers a major source of emissions to be one that emits more than 10 tons per year of a hazardous air pollutant, and it allows the Administrator to establish a lesser quantity as the major source threshold, based on the potency of the air pollutant or other relevant factors.

Once the source categories for hazardous air pollutants are identified, Section 112 establishes a presumption in favor of regulation of the designated pollutants; it requires regulation unless EPA or a petitioner is able to show "that there is adequate data on the health and environmental effects of the substance to determine that emissions, ambient concentrations, bioaccumulation or deposition of the substance may not reasonably be anticipated to cause any adverse effects to human health or adverse environ-mental effects."

## Path 3. Regulating GHGs through Sections 111 as Designated Air Pollutants

Given the difficulties in following the first two paths, much of the attention, including EPA's, has been on the third path. The term "designated pollutant" is a catch-all phrase for any air pollutant that isn't either a criteria air pollutant under Section 108 or a toxic air pollutant under Section 112. Examples of these include fluorides from phosphate fertilizer manufacturing or primary aluminum reduction, or sulfuric acid mist from sulfuric acid plants.

### *Importance of Section 111*

The authority to regulate such pollutants is Section 111.[25] Section 111 establishes New Source Performance Standards (NSPS), which are emission limitations imposed on designated categories of major new (or substantially modified) stationary sources of air pollution. A new source is subject to NSPS regardless of its location or ambient air conditions.[26]

Section 111 provides authority for EPA to impose performance standards on stationary sources— directly in the case of new (or modified) sources, and through the states in the case of existing sources (Section 111(d)). The authority to impose performance standards on new and modified sources refers to any category of sources that the Administrator judges "causes, or contributes significantly to, air pollution which may reasonably be anticipated to endanger public health or welfare" (Sec. 111(b)(1)(A)). In establishing these standards, the Administrator has the flexibility to "distinguish among classes, types, and sizes within categories of new sources" (Sec. 111(b)(2)).

The performance standards themselves are to reflect "the degree of emission limitation achievable through the application of the best system of emission reduction which (taking into account the cost of achieving such reduction and any nonair quality health and environmental impact and energy requirements) the Administrator determines has been adequately demonstrated" (Sec. 111(a)(1)). Both the Administrator and the individual states have the authority to enforce the NSPS.

### Controlling GHG through Section 111

Section 111 appears to provide a strong basis for EPA to establish a traditional regulatory approach to controlling greenhouse gas emissions from large stationary sources. As noted, the section gives EPA considerable flexibility with respect to the source categories regulated, the size of the sources regulated, the particular greenhouse gases regulated, along with the timing and phasing in of regulations. This flexibility extends to the stringency of the regulations with respect to costs, and secondary effects, such as nonair quality, heath and environmental impacts, along with energy requirements. This flexibility is encompassed within the Administrator's authority to determine what control systems she determines have been "adequately demonstrated." As discussed later, this determination has been used to authorize control regimes that extended beyond the merely commercially available to those technologies that have only been demonstrated, and thus are considered by many to have been "technology-forcing."

In sum, Section 111 has several advantages in considering greenhouse gas controls including that it (1) has flexibility with respect to the size of the source controlled (Section 111 (b)(2)), (2) can prioritize its schedule of performance standards (Section 111(f)(2)), (3) can consider costs and other factors in making determinations, and (4) has discretion with respect to determining technology that has been adequately demonstrated. Essentially, using Section 111, EPA can determine who gets controlled, when they get controlled, how much they get controlled, and at what price.

## Going off the Beaten Path: Regulating under Section 115 or Title VI

### Section 115. International Pollution

On the face of it, Section 115 would appear the ideal provision to address the global issue of climate change. It is focused on international problems and

Climate Change      89

has unique international triggers. Specifically, Section 115 could be invoked by EPA on one of two bases.

First, EPA could act if it receives reports, surveys, or studies from "any duly constituted international agency" that gives EPA:

> reason to believe that any air pollutant or pollutants emitted in the United States cause or contribute to air pollution which may reasonably be anticipated to endanger public health or welfare in a foreign country....[27]

Unlike the endangerment triggers under other sections of the Act, the endangerment finding under Section 115 refers to international effects based on data from internationally recognized sources. Many would argue that reports by the Intergovernmental Panel on Climate Change (IPCC) would fit this requirement. A United Nations body, created by the World Meteorological Organization and United Nations Environment Programme, the group and its results are referenced by EPA in its ANPR and its endangerment finding under Section 202, announced December 7, 2009.

Second, in addition to a unique international endangerment trigger, Section 115 can be invoked without any EPA endangerment finding at all. Specifically, EPA is directed to act "whenever the Secretary of State requests him to do so with respect to such pollution [that endangers public health or welfare in a foreign country] which the Secretary of State alleges is of such a nature...." (Section 115(a)). Thus, an allegation by the Secretary of State is sufficient cause for EPA to act.

The action called for under Section 115 is implemented through Section 11 0(a)(2)(H)(ii) that requires states to revise their SIPs to prevent or eliminate the endangerment identified. Apparently, based on this reference to SIPs, EPA states in its ANPR that Section 115 could only be exercised if EPA were to promulgate a NAAQS for greenhouse gases.[28] However, this is arguable. Section 1 10(a)(2)(H)(ii) states that SIPs must be crafted to provide for revisions:

> ...whenever the Administrator finds on the basis of information available to the Administrator that the plan is substantially inadequate to attain the national ambient air quality standard which it implements *or to otherwise comply with any additional requirements established under this Act. [emphasis added]*

In their article arguing in favor of using Section 115 to address climate change, Martella and Paulson state their opposition to EPA's blanket assertion that a greenhouse gas NAAQS would be necessary to invoke Section 115:

... based on the plain language of the statute, however, this is unlikely to have been what Congress intended. Section 115 is not in any way limited to criteria pollutants. In fact, the opposite is true. It applies specifically to "any air pollution." Clean Air Act Section 11 0(a)(2)(H)(ii) makes it clear that SIP must provide for the revision of the plan not only when the plan is inadequate to attain a NAAQS, but also to otherwise comply with any additional requirements, such as a revision required by Section 115.[29] [footnotes omitted]

The above actions are prefaced on a condition of reciprocity; Section 115 applies "only to a foreign country which the Administrator determines has given the United States essentially the same rights with respect to the prevention or control of air pollution occurring in that country as is given that country by this section." (Section 115(c)) EPA notes in its ANPR that reciprocity with one or more affected countries may be sufficient to trigger Section 115.[30] Many countries currently attempting to comply with the Kyoto Protocol, such as the European Union, could argue that their efforts to reduce greenhouse gases are being hindered by absent or inadequate U.S. controls. Such countries could argue they meet the criteria under Section 115(c) with respect to reciprocity and point to international studies supporting their position. Secondly, countries at substantial risk from climate change, such as low-lying island countries, could argue endangerment from the lack of U.S. action. Thirdly, countries that only contribute a *de minimis* level of emissions, such as virtually all of Africa, could argue that their low emissions meet the criterion for U.S. action.

Subject to the limitations of the SIP process, EPA notes that Section 115 would provide it with some flexibility in program design. Martella and Paulson take a much more expansive view of the flexibility available, arguing:

While designating SIPs as the implementation vehicle, Section 115 otherwise does not impose strictures on the contours and requirements of any prospective program(s) to reduce greenhouse gas emissions.... A Section 115-based program could therefore include model thresholds and source categories set by EPA, similar to the Northeast Ozone Transport.

Additionally, EPA could develop a holistic model plan to be implemented by the states. Multiple model approaches also could be presented to the states allowing each state to pick the most appropriate solution for its particular mix of greenhouse gas sources....
Additionally, Section 115 provides a mechanism to limit the scope of the program in terms of the sources....[31]

Because EPA asserts that invoking Section 115 would require a greenhouse gas NAAQS, the action would also invoke NSR under Part C and Title V permitting requirements. One of Martella and Paulson's primary arguments in favor of Section 115 is their belief that Section 115's unique endangerment requirements (or no endangerment requirement if the Secretary of State alleges endangerment) should not trigger PSD-NSR or Title V permitting requirements.[32]

Finally, it should be noted that Section 115 has never been implemented, and many countries would prefer a negotiated settlement on climate change, rather than this approach.

## *Title VI. Stratospheric Ozone Protection*

Added to the Clean Air Act in 1990, Title VI is the country's implementing legislation for the Montreal Protocol and succeeding agreements to address ozone depletion by human-made substances. Some of the substances that deplete the ozone layer also contribute to climate change (e.g., CFCs, HCFCs). In addition, some substances chosen as substitutes for ozone depleting chemicals are themselves greenhouse gases (e.g., HFC-134a, PFCs). Finally, the process of making acceptable substitutes for more powerful ozone-depleting chemicals (e.g., HCFC-22) produces greenhouse gases as a byproduct of production (e.g., HFC-23).

Beyond these chemical relationships, there is continuing research on the atmospheric relationship between the stratosphere (and the ozone layer) and climate change.

There are two provisions of Title VI that could be used to address greenhouse gas emission under certain conditions. They are discussed below.

## Section 612. Safe Alternatives Policy

As noted above, some substitutes for ozone-depleting substances are greenhouse gases, such as HFCs and PFCs. Section 612 authorizes EPA to the maximum extent practicable, to identify substitutes for ozone-depleting chemicals that reduce overall risks to human health and the environment.

Specifically, Section 612(c) requires the EPA to make it unlawful to replace an ozone-depleting substance with any substitute substance which EPA determines "may present adverse effects to human health or the environment" where EPA has identified an available, less harmful substitute. The resulting program is called the Significant New Alternatives Policy (SNAP). With appropriate substitutes identified, SNAP could be used to reduce emissions of HFCs and PFCs without invoking any other provisions of the CAA.

## Section 615. Authority of Administrator

Like Section 115, Section 615 is potentially a powerful mechanism to control greenhouse gas emissions under certain circumstances. Like Section 115, it has a unique endangerment finding requirement and even broader discretionary authority for EPA to respond. Section 615 states:

> If, in the Administrator's judgment, any substance, practice, process, or activity may reasonably be anticipated to affect the stratosphere, especially ozone in the stratosphere, and such effect may reasonably be anticipated to endanger public health or welfare, the Administrator shall promptly promulgate regulations respecting the control of such substance, practice, process or activity, and shall submit notice of the proposal and promulgation of such regulation to the Congress.

Invoking Section 615 in the case of greenhouse gases would involve a two-part judgment by the EPA: First, that greenhouse gases may reasonably be anticipated to affect the stratosphere (particularly the ozone layer) and, second, that the effect on the stratosphere may reasonably be anticipated to endanger public health or welfare. In its ANPR, EPA determined that it was beyond the scope of its ANPR to assess and analyze the available scientific information on the effects of greenhouse gases on the stratosphere.

If EPA were to judge the scientific data adequate to meet the two-part test, the authority available would be broad and deep. As stated by EPA in its ANPR: "... depending on the nature of any finding made, section 615 authority may be broad enough to establish a cap-and-trade program for the substance, practice, process or activity covered by the finding.... "[33]

## POTENTIAL CONTROL APPROACHES FOR STATIONARY SOURCES

In its Technical Support Document for its ANPR, EPA took a narrow view of the alternatives available to it in imposing greenhouse gas performance standards.[34] For existing electric generating sources, the EPA focused on incremental improvements in the heat rates of existing units through options that "are well known in the industry" with an overall improvement in efficiency likely to be less than 5%. For new electric generating sources, EPA noted the availability of more efficient supercritical coal units, the future availability of ultra-supercritical units, and the possibility of limited biomass co-firing.

Continuing along this line of reasoning, EPA also suggested that it could develop regulations that anticipate future technology. For example, a phase-in approach to applying $CO_2$ standards to powerplants would be to mandate that "carbon-ready" generating technology be required for new construction. The objective would be to anticipate the widespread need for some form of carbon capture technology in the future by preparing for it with compatible fossil-fuel combustion technology now. The technology most discussed is integrated-gasification, combined-cycle (IGCC). As noted earlier, EPA is considering this option with respect to the *Desert Rock* PSDNSR permit reconsideration. With respect to some of the carbon capture technology under development, IGCC has certain advantages over pulverized coal technology. However, just how much IGCC is "carbon ready" is subject to debate. EPA states in its ANPR that it believes such a staged approach is available to it under section 111:

> EPA believes that section 111 may be used to set both single-phase performance standards based upon current technology and to set two-phased or multi-phased standards with more stringent limits in future years. Future-year limits may permissibly be based on technologies that, at the time of the rulemaking, we find adequately demonstrated to be available for use at some specified future date.[35]

The technical support document does not mention some more aggressive options. These include a fuel-neutral standard or a technology-based standard. For example, for carbon dioxide emissions from a newly-constructed powerplant, a fuel-neutral standard could follow the example set by the 1997 and 2005 NOx NSPS and the 2005 NOx NSPS for modified existing sources. Under those regulations, the NOx emissions standard is the same, regardless of the fuel burned—solid, liquid, or gaseous.[36] This standard is much more expensive for coal-fired facilities to comply with than for natural-gas-fired facilities, thus encouraging the lower-carbon gas-fired technologies. Likewise, EPA could choose to set a newly-constructed powerplant standard based on the performance of natural gas burned in a combined-cycle configuration – the fuel and technology of choice for construction of new powerplants for the last two decades. If EPA wanted to encourage the rollover of the existing coal-fired powerplant fleet to natural gas, nuclear, or renewable sources, it could apply a fuel-neutral standard to modified sources as well. For example, a $CO_2$ emission standard of 0.8 lb. per kilowatt-hour output could be met by a new natural gas-fired, combined-cycle facility, as well as any non-emitting generating technology, such as nuclear power or renewables. In contrast, the standard would require a 60% reduction in emissions from a new coal-fired facility – forcing the development of a carbon control technology, such as carbon capture and storage (CCS), in order for a new coal-fired facility to be built or modified.

The viability of these options, or even more aggressive technology-forcing standards, would depend on how EPA determined whether a technology had been "adequately-demonstrated" and the seriousness of its costs and energy requirements. As discussed below, EPA has used the NSPS to encourage the installation of pollution control equipment on powerplants, even while the equipment's development status was still being debated.

## Forcing Commercialization of Technology through a Regulatory Requirement: An Example from the $SO_2$ New Source Performance Standards

It is an understatement to say that the new source performance standards promulgated by the EPA were technology-forcing. Electric utilities went from having no scrubbers on their generating units to incorporating very complex chemical processes. Chemical plants and refineries had scrubbing systems that were a few feet in diameter, but not the 30- to 40-foot diameters required by the utility industry. Utilities had dealt with hot flue gases, but not with saturated flue gases that contained all sorts of contaminants. Industry, and the US EPA, has always looked upon new source performance standards as technology-forcing, because they force the development of new technologies in order to satisfy emissions requirements.[37]

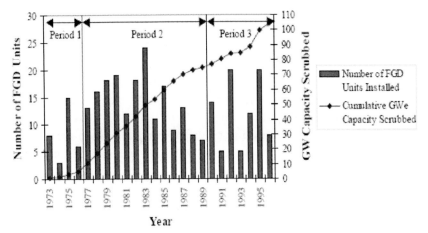

Source: Adapted by Taylor from Soud (1994). See Margaret R. Taylor, op. cit., 74.
Note: Numbers are archival through June 1994, then projected for 1994-96.

Figure 1. Number of FGD Units and Cumulative Gigawatt (GW) Capacity of FGD Units: 1973-1996.

The most direct method to encourage adoption of carbon capture technology would be to mandate it. Mandating a performance standard on stationary sources is not a new idea: The process of forcing the development of emission controls on coal-fired powerplants is illustrated by the 1971 and 1978 $SO_2$ NSPS for coal-fired electric generating plants. As noted earlier, the Clean Air Act states that NSPS should reflect "the degree of emission limitation achievable through the application of the best system of emission reduction which (taking into account the cost of achieving such reductions and any non-air quality health and environmental impact and energy requirements) the Administrator determines has been adequately demonstrated."[38] In promulgating its first utility $SO_2$ NSPS in 1971, EPA determined that a 1.2 pound of $SO_2$ per million Btu of heat input performance standard met the criteria of Sec. 111—a standard that required, on average, a 70% reduction in new powerplant emissions, and could be met by low- sulfur coal that was available in both the eastern and western parts of the United States, or by the use of emerging flue gas desulfurization (FGD) devices.[39]

At the time the 1971 Utility $SO_2$ NSPS was promulgated, there was only one FGD vendor (Combustion Engineering) and only three commercial FGD units in operation—one of which would be retired by the end of the year.[40] The number of units and vendors would increase rapidly, not only because of the NSPS, but also because of the promulgation of the $SO_2$ NAAQS, the 1973

Supreme Court decision preventing significant deterioration of pristine areas,[41] and state requirements for stringent $SO_2$ controls, which opened up a market for retrofits of existing coal- fired facilities in addition to the NSPS focus on new facilities. Indeed, most of the growth in FGD installations during the early and mid-1970s was in retrofits. Taylor estimates that between 1973 and 1976, 72% of the FGD market was in retrofits.[42] By 1977, there were 14 vendors offering full-scale commercial FGD installation.[43]

However, despite this growth, only 10% of the new coal-fired facilities constructed between 1973 and 1976 had FGD installations. In addition, the early performance of these devices was not brilliant.[44] In 1974, American Electric Power (AEP) spearheaded an ad campaign to have EPA reject FGD devices as "too unreliable, too impractical for electric utility use" in favor of tall stacks, supplementary controls, and low-sulfur western coal.[45] This effort was ultimately unsuccessful as the Congress chose to modify the NSPS requirements for coal-fired electric generators in 1977 by adding a "percentage reduction" requirement. As promulgated in 1979, the revised $SO_2$ NSPS retained the 1971 performance standard but added a requirement for a 70%-90% reduction in emissions, depending on the sulfur content of the coal.[46] At the time, this requirement could be met only through use of an FGD device. The effect of the "scrubber requirement" is clear from the data provided in **Figure 1**. Based on their analysis of FGD development, Taylor, Rubin, and Hounshell state the importance of demand-pull instruments:

> Results indicate that: regulation and the anticipation of regulation stimulate invention; technology-push instruments appear to be less effective at prompting invention than demand- pull instruments; and regulatory stringency focuses inventive activity along certain technology pathways.[47]

That government policy could force the development of a technology through creating a market should not suggest that the government was limited to that role, or that the process was smooth or seamless. On the latter point, Shattuck, et al., summarize the early years of FGD development as follows:

> The Standards of Performance for New Sources are technology-forcing, and for the utility industry they forced the development of a technology that had never been installed on facilities the size of utility plants. That technology had to be developed, and a number of installations completed in a short period of time. The US EPA continued to force technology through the promulgation of successive

regulations. The development of the equipment was not an easy process. What may have appeared to be the simple application of an equipment item from one industry to another often turned out to be fraught with unforeseen challenges.[48]

The example indicates that technology-forcing regulations can be effective in pulling technology into the market—even when there remain some operational difficulties for that technology. The difference for carbon capture technology is that for long-term widespread development, a new infrastructure of pipelines and storage sites may be necessary in addition to effective carbon capture technology.[49] In the short-term, suitable alternatives, such as enhanced oil recovery needs and in-situ geologic storage, may be available to support early commercialization projects without the need for an integrated transport and storage system. Likewise, with economics more favorable for new facilities than for retrofits, concentrating on using new construction to introduce carbon capture technology might be one path to widespread commercialization. As an entry point to carbon capture deployment, a regulatory approach such as NSPS may represent a first step, as suggested by the $SO_2$ NSPS example above.

## Potential for Cap-and-Trade

Whether EPA can set up a cap-and-trade program under the Clean Air Act is the subject of considerable debate in the literature.[50] Much of the debate surrounds the provisions of Section 111(d). However, there are other authorities in the Act that might serve as a basis for a EPA- coordinated cap-and-trade program.

### Potential Under Section 111

EPA, along with other commenters, has linked the potential effectiveness of Section 111(d) to whether it can be interpreted to allow a cap-and-trade program for $CO_2$ . As stated by EPA: "EPA also believes that because of the potential cost savings, it might be possible for the Agency to consider deeper reductions through a cap-and-trade program that allowed trading among sources in various source categories relative to other systems of emissions reduction."[51] As noted, Section 111 explicitly allows EPA to take cost into consideration in developing performance standards. Whether that consideration could justify a trading program across different greenhouse gases, and across different source categories with different best available systems of emissions reduction is not known. A lead author of the winning brief in *Massachusetts v. EPA* makes a case against such authority:

> Numerous parties have argued that section 111 does not authorize the creation of a cap-andtrade program. Among other things, section 111 (h) provides a contingency plan in the event performance standards are "not feasible" to implement. In that case, section 111(h) gives EPA the authority to "promulgate a design, equipment, work practice, or operational standard, or combination thereof, which reflects the best technological system of continuous emissions reduction which ... the Administrator determines has been adequately demonstrated." 42 U.S.C. Section 741 1(h)(1). One of the ways a performance standard might prove "not feasible" is if "a pollutant or pollutants cannot be emitted through a conveyance designed and constructed to emit or capture such pollutants." 42 U.S.C. 7411 (h)(2)(A). Clearly, Congress thought the most likely scenario under section 111 was for pollutants to be "emitted through a conveyance designed and constructed to emit or capture such pollutant[s]" – an assumption at odds with the operation of a trading program. Other aspects of section 111 also point away from the creation of a trading program under this provision [reference omitted].[52]

In sum, whether this authority can be expanded to creating a comprehensive cap-and-trade program is under debate. Focused on existing sources, EPA used Sec. 111(d) to justify its promulgated rule (now vacated) to reduce mercury emissions from powerplants. Although some have argued that the court decision in this case repudiated EPA's reasoning, the case was actually not decided on the basis of Section 111(d).[53]

### Potential under Other Sections

Three other sections of the Act, (Sections 110, 115, and 615) might also be considered as possible authority for establishing an economy-wide cap-and-trade program for GHG emissions, although each has its own weaknesses. Section 110 of the Act establishes requirements for State Implementation Plans (SIPs). While primarily designed to demonstrate how a state with nonattainment areas will bring those areas into attainment with NAAQS, the section also contains language that might serve as the basis for the use of broader GHG regulatory tools once emission standards were issued under any section of the Act. Specifically, Section 11 0(a)(2)(A) says that each SIP shall

## Climate Change

... include enforceable emission limitations and other control measures, means, or techniques (including economic incentives such as fees, marketable permits, and auctions of emissions rights), as well as schedules and timetables for compliance, as may be necessary or appropriate to meet the applicable requirements of this Act ....

The predicate is that there must first be an applicable requirement under the Act. Thus, Section 110 would not be an authority that EPA could use to *initiate* regulation of GHGs. Also, although the section mentions economic incentives, marketable permits, and auctions, it is not clear that such authority could be used for economy-wide control measures. The precedents for the authority's use that EPA cited in the ANPR, for example, included such regulations as the NOx SIP call, which established a cap-and-trade program for powerplant emissions of NOx, and the Clean Air Interstate Rule, which also allowed trading of emission allowances by powerplants.

As stated in the ANPR:

EPA has often incorporated market-oriented emissions trading elements into the more traditional performance standard approach for mobile and stationary sources. Coupling market-oriented provisions with performance standards provides some of the cost advantages and market flexibility of market-oriented solutions while also directly incentivizing technology innovation within the particular sector, as discussed below. For example, performance standards for mobile sources under Title II have for many years been coupled with averaging, banking and trading provisions within a subsector. In general, averaging allows covered parties to meet their emissions obligation on a fleet- or unit-wide basis rather than requiring each vehicle or unit to directly comply. Banking provides direct incentives for additional reductions by giving credit for overcompliance; these credits can be used toward future compliance obligations and, as such, allow manufacturers to put technology improvements in place when they are ready for market, rather than being forced to adhere to a strict regulatory schedule that may or may not conform to industry or company developments. Allowing trading of excess emission reductions with other covered parties provides an incentive for reducing emissions beyond what is required.[54]

The two other possible authorities for a cap-and-trade program, Sections 115 and Section 615, have never been used to control any pollutant, much less to establish a cap-and-trade program. Assuming Section 115 could be invoked without a supporting NAAQS, there might be sufficient flexibility to institute a cap-and-trade program. The program would have to be created by each state under Section 110 to comply with EPA-determined state GHG emission caps in response to Section 115. Because it would function through Section 110, EPA could not impose a cap-andtrade system on the states; rather, the states would have to voluntarily agree to cooperate in a EPA-coordinated cap-and-trade scheme.

As noted earlier, if Section 615 could be successfully triggered by the science, EPA's discretion in setting up a regulatory scheme would be substantial. As stated by EPA in its ANPR: "... depending on the nature of any finding made, section 615 authority may be broad enough to establish a cap-and-trade program for the substance, practice, process or activity covered by the finding.... "[55]

## IMPLEMENTATION ISSUES

### New Source Review

Any new or modified facility emitting (or potentially emitting) over 250 tons of any regulated pollutant must undergo preconstruction review and permitting, including the installation of Best Available Control Technology (BACT), except those pollutants regulated under Sections 112 (hazardous air pollutants) and 211(o) (renewable fuels). New sources under the Prevention of Significant Deterioration provisions of Part C (PSD-NSR) must undergo preconstruction review and must install BACT as the minimum level of control.[56] State permitting agencies determine BACT on a case-by-case basis, taking into account energy, environmental, and economic impacts. BACT cannot be less stringent than the federal NSPS, but it can be more so. More stringent controls can be required if modeling indicates that BACT is insufficient to avoid violating PSD emission limitations, or the NAAQS itself.

PSD-NSR is required for any pollutant "subject to regulation" under the Clean Air Act, but there are varying interpretations of what the phrase "subject to regulation" means. Environmental groups have argued that $CO_2$ is already subject to regulation because utilities are required under Section 821 of the Clean Air Act Amendments of 1990 to monitor and report $CO_2$ emissions to EPA. Others argue that an endangerment finding would make GHGs subject to regulation, and, therefore, trigger PSD-NSR requirements for new sources. In its endangerment finding and in the preamble to the proposed Greenhouse Gas

Climate Change 101

Tailoring Rule, EPA noted its current interpretation of the law is that a final positive endangerment finding for motor vehicles under Section 202 would not *per se* make greenhouse gas emissions subject to PSD-NSR.[57] Rather, as stated in a September 30, 2009 proposal, "Although several possible triggering events may be considered ..., the latest of these events would be the one that applies under EPA's current interpretation: a nationwide rule controlling or limiting GHG emissions."[58] However, the interpretive memorandum on which this conclusion is based,[59] issued in December 2008, is currently under review by the new Administration.[60]

## Issue of Case-by-Case BACT Determinations

Two aspects of the New Source Review provision create potential difficulties in using the CAA to control greenhouse gases. First, as noted earlier, PSD-NSR has specified thresholds for triggering its provisions: a "major emitting facility is generally defined as emitting or having the potential to emit either 100 tons or 250 tons annually of a regulated pollutant (Sec. 169(1)).[61] With respect to greenhouse gases, this is a fairly low threshold. By comparison, several bills introduced in the 110th Congress set thresholds for inclusion in the reduction program at 10,000 metric tons annually, and the Waxman-Markey bill (H.R. 2454) generally uses 25,000 tons as a regulatory threshold.

The second administrative issue for PSD-NSR is the requirement that BACT be determined on a case-by-case basis. Combined with a 100-ton or 250-ton threshold, this could mean a massive increase in state-determinations of BACT: the resulting increased permit activity would be at least two orders of magnitude, according to EPA (discussed below).

On this second issue, it should be noted that several commenters believe this would not be a major problem (unless a cap-and-trade program is implemented). As stated by the Institute for Policy Integrity:

> Since including GHGs in the PSD program may greatly expand the number of permits issued, making case-by-case determinations for each individual source may stretch the resources of EPA and state permitting authorities. Moreover, traditional technological controls may not exist for every GHG emitted by every regulated facility. However, there is flexibility in the statute to resolve these problems.

Though BACT determinations are generally to be made on a case-by-case basis, the D.C. Circuit recognized in *Alabama Power* that exceptions can be made if "case-by-case determinations would, as a practical matter, prevent the agency from carrying out the mission assigned to it by Congress." The development of "presumptive BACT" determinations should be permissible and may help streamline the permitting process [footnote omitted]."[62]

In addition, assuming PSD is triggered by regulation under Section 111, the BACT requirements may be identical to the NSPS determinations under Section 111. It is also likely that most small sources would not have an NSPS as EPA applied its discretion under Section 111 in determining the most cost-effective emissions reductions. With no NSPS floor for a BACT determination, it is possible that NSR requirements for sources not covered under Section 111 could be quite lax.

In addition, EPA has proposed to address the threshold problem in a Greenhouse Gas Tailoring Rule. The proposal would set a 25,000-ton $CO_2$-equivalent threshold for new source permitting under the PSD-NSR program, and would establish a significance level for determining major modifications of 10,000-25,000 tons of increased annual emissions.

## Title V and the Size Threshold

In the ANPR, EPA discussed the possibility that an endangerment finding and subsequent regulation of GHGs as air pollutants under any section of the Act could trigger Title V permit requirements, and that all facilities that have the potential to emit a GHG pollutant in amounts of 100 tons per year or more would be required to obtain permits. Under this reasoning, the regulation of $CO_2$ from motor vehicles under Section 202, for example, could lead to Title V permit requirements for $CO_2$ from powerplants and other sources. In the ANPR, the agency stated:

> Using available data, which we acknowledge are limited, and engineering judgment in a manner similar to what was done for PSD, EPA estimates that more than 550,000 additional sources would require Title V permits, as compared to the current universe of about 15,000– 16,000 Title V sources. If actually implemented, this would be more than a tenfold increase, and many of the newly subject

sources would be in categories not traditionally regulated by Title V, such as large residential and commercial buildings.[63]

In the preamble to its September 30, 2009 Tailoring Rule, EPA increased its estimate to more than 6 million sources potentially subject to Title V if the threshold remains at 100 tons per year of emissions.[64]

Thus, like PSD-NSR, a major complication that Title V introduces is the potential for very small sources of greenhouse gases to need permits in order to operate. Furthermore, Title V requires that covered entities pay fees established by the permitting authority, and that the total fees be sufficient to cover the costs of running the permit program.

The potential for increased permitting activity has led to speculation on its potential extent. For example, some agricultural interests have spun the possibility that Title V could be invoked for emissions from agricultural activities and the requirement for permit fees into something they refer to as the "cow tax." On November 18, 2008, for example, Cattle Network stated "EPA Proposes 'Cow Tax.'" The article even generated specific amounts for the "tax": $175 per dairy cow and $87.50 per beef cow.[65] EPA says that it has no plans to regulate agricultural activities' GHG emissions. Indeed, the agency currently exempts most major agricultural sources from any Clean Air Act controls on conventional air pollutants under an arrangement known as the Air Compliance Agreement.[66] Thus, it would seem unlikely that the agency would now make a priority of subjecting small agricultural sources to GHG requirements.

However, the need to deal with the size issue has been noted by EPA and other commenters. Alternatives to lessen the extent and cost of these provisions fall into three categories: (1) legal or regulatory interpretations that increase EPA's flexibility to determine what sources would need permits and when; (2) the expanded use of general permits; or (3) interpretation of different endangerment findings to exclude Title V and/or PSD-NSR.

### *Legal or Regulatory Interpretations that Increase Flexibility*

EPA noted two possible legal theories under which it could avoid imposing PSD-NSR or Title V permitting requirements on small sources. Under "the judicial doctrine of administrative necessity," the agency stated that it might be able "to craft relief in the form of narrowed source coverage, exemptions, streamlined approaches or procedures, or a delay of deadlines."[67] The agency also stated that in rare cases, the courts will apply statutory provisions in a manner other than that indicated by the plain meaning, if "absurd, futile, strange, or indeterminate results" would be produced by literal application.

Following up on these interpretations, EPA proposed a Title V and PSD-NSR permitting threshold of 25,000 tons per year of $CO_2$-equivalents ($CO_2$-e) on September 30, 2009. The preamble to the rule states:

> This proposal is necessary because EPA expects soon to promulgate regulations under the CAA to control GHG emissions and, as a result, trigger PSD and title V applicability requirements for GHG emissions. If PSD and title V requirements apply at the applicability levels provided under the CAA, state permitting authorities would be paralyzed by permit applications in numbers that are orders of magnitude greater than their current administrative resources could accommodate. On the basis of the legal doctrines of "absurd results" and "administrative necessity," this proposed rule would phase in the applicability thresholds for both the PSD and title V programs for sources of GHG emissions.[68]

While setting a new threshold for permitting, the proposal does not exempt smaller sources. Rather, the agency and state-permitting authorities, within five years of the rule's promulgation, would conduct a study of the permitting authorities' ability to administer the programs going forward and, within a year of the study's completion, would conduct a rulemaking for the second phase of the program. The study might confirm the threshold, revise it, or establish other streamlining techniques for subsequent permitting activity.

Where EPA has the authority, such as under Section 111, it will almost certainly focus on the large sources first. As noted in the introduction, when it comes to stationary sources, size matters. Twenty-eight percent of the country's GHGs comes from an Energy Information Administration (EIA) estimated 670 coal-fired electric powerplants. Farms, by contrast, number more than 2 million, and emit less than 4% of total GHGs. Methane (CH4) provides another interesting contrast in potential priorities: about 1.8% of GHG emissions, in the form of methane, are generated by 1,800 landfills; a slightly larger amount (2.4%) is emitted by roughly a million cattle and swine operations. As stated by the Institute for Policy Integrity:

> Courts grant agencies much more leeway in deferring full implementation of a statute than in creating permanent exemptions. Invoking the doctrine of administrative necessity, EPA should be able to justify expanding NSR permit applicability to the largest sources

first, and then gradually including smaller sources. The timeline set for phasing in smaller sources could not take longer than reasonably necessary given EPA's administrative burdens, but EPA will have a good deal of discretion to determine its own resources and capability [footnotes omitted].[69]

A second means of reducing the administrative burden is to increase the effective size of an affected source by defining "potential to emit" in terms of potential actual emissions. In particular, EPA suggested in its ANPR that determining the potential to emit in terms of actual usage instead of maximum potential could have some benefit in some cases. For example, if a small boiler's potential to emit was based on actual usage of 1000 hours a year, instead of continuous potential usage (8760 hours), many fewer boilers would be subject to NSR.[70]

### *General Permits*

Perhaps the most straightforward method of reducing administrative burden is for EPA to adopt a general permit scheme for PSD-NSR and Title V. For categories with numerous similar sources of emissions, the Clean Air Act provides in Section 5 04(d) that the permitting authority—be it EPA or a delegated state agency—may issue a "general permit" covering all sources in the category. This provision substantially reduces the administrative burden of issuing permits, allowing notice and opportunity for public hearing on the category as a whole and the provisions of the general permit, rather than requiring the same for each individual source. General permits have been widely used by the agency under the Clean Water Act, and are used by about half the states for control of various air pollution sources. Thus, there is precedent for their use in a Clean Air Act greenhouse gas control program for multiple, relatively minor sources of emissions.

A general permit does not relieve the permittee from filing a permit application or from complying with permit conditions, which would include some sort of monitoring and reporting requirements. But a permit application for a general permit can be relatively simple, and since there are few costs to issuing the permit, permit fees, which are required by Section 502(b) to cover the reasonable costs of the permit program, but are to be utilized only to cover such costs, would be relatively low. A sampling of states using general permit fees for other types of air pollutants found fees ranging from $100 to $350 per permittee.

Such an approach may also be available to small sources potentially caught under PSD-NSR. Both EPA in the ANPR and the Institute for Policy Integrity provide arguments for PSD-NSR general permits for small sources to avoid absurd results or respond to administrative necessity.[71]

## Section 304. Citizen Suits

If an endangerment finding triggered emissions standards or limitations under the CAA (e.g., Section 111, Part C), it would also bring into play Section 304, Citizen Suits. Section 304 allows any person to commence a civil action against any other person (including government entities and instrumentalities) for violation of an emissions standard or limitation under the Act. It also provides for suits against EPA for failing to perform a nondiscretionary act or duty. Most specifically, Section 304 provides for suits

> against any person who proposes to construct or constructs any new or modified major emitting facility without a permit required under part C of title I (relating to significant deterioration of air quality) or part D of title I (relating to non-attainment) or who is alleged to have violated (if there is evidence that the alleged violation has been repeated) or to be in violation of condition of such permit.[72]

Citizen suits have been widely used by environmental groups to force the Administrator to undertake nondiscretionary duties and to enforce the Act's requirements against emitting facilities. Should the agency fail to move forward with GHG standards following an endangerment finding, suits seeking to force action would almost certainly be filed.

## CONCLUSION

The current debate on the appropriateness of using the Clean Air Act to regulate greenhouse gas emissions is not the first such debate that has occurred when a new environmental challenge has been directed at the Act. During the 1980s, suggestions were made that acid rain and/or stratospheric ozone depletion could be addressed via then-existing provisions, rather than by new Amendments. For example, in 1985, the CRS stated the following with respect to addressing acid rain through the existing Clean Air Act:

# Climate Change

107

*Various Clean Air Act provisions could be used to address acid precipitation,* including issuing more stringent secondary ambient air quality standards, setting a sulfate standard, and enforcing $SO_2$ reductions more vigorously. (a) Typically, however, such actions require a demonstration of cause-effect relationship that has not been obtained, at least in the view of many policymakers; and/or they require actions under peripherally related provisions such as visibility protection—which are already subject to controversy on their own right. (b) Any such actions would likely be expensive, both in resources and in political/administrative capital. (c) *Program administrators have therefore said they will not use the Clean Air Act aggressively and innovatively to combat acid precipitation without an explicit Congressional mandate and/or compelling new evidence linking specific damages to specific pollutants* [emphasis in original].[73]

In both cases, the Congress moved to add new Titles to the Act (Title IV to address acid rain, and Title VI to address stratospheric ozone depletion). In the case of Title IV, a new market-based approach to reducing pollutants was introduced to implement a statutory reduction requirement (i.e., the $SO_2$ emissions cap) in hope that the cost would be optimized. The result was so successful that it was used by states and EPA to begin addressing interstate transport of smog (i.e., the NOx SIP Call) and has been suggested by some as the optimal approach to controlling greenhouse gases.

However, controlling greenhouse gases is a substantially more complex environmental, technical, economic, and social issue than either acid rain or stratospheric ozone depletion are. It is possible that one size does not fit all in this debate. Some sources may not respond significantly to a market-based approach because they are not particularly price-sensitive. Others may be too small or dispersed to include. For example, the European Union's market-based approach covers only about 40% of the EU's emissions. Other instruments are used to address difficult sectors, such as transportation.

Thus, initiatives to use the current Clean Air Act could be designed as a substitute for what is perceived by some as a protracted congressional debate, or as a complementary effort to address sources or gases that a future market-based system may choose to exclude from its provisions. As summarized in 2008 by Lisa Heinzerling in testimony to the Subcommittee on Energy and Air Quality of the House Energy and Commerce Committee:

the Clean Air Act contains numerous provisions that might be used to regulate greenhouse gases. The advantages of using these provisions include: they can be deployed now; they use regulatory strategies that are familiar to, indeed are the bread and butter work of, the Environmental Protection Agency; they call for regulation of numerous and diverse sources and thus, taken as a group, they have an inherent fairness to them; they do not pose unusual enforcement difficulties or untoward administrative burdens.

There are also disadvantages to using existing Clean Air Act provisions to address climate change. Most of the provisions do not have statutory deadlines.... To the extent one favors cap-and-trade as a regulatory mechanism for addressing climate change, one might worry about the lack of clear authority for such a scheme under the existing statute. The NAAQS program is an ungainly framework for regulating globally harmful pollutants. PSD requirements are triggered for sources that are "large" when it comes to conventional pollution but "small" from the perspective of global pollutants.[74]

A final endangerment finding presents EPA with many options. However, the ultimate decision on what the Nation's greenhouse gas policy should be rests with the Congress. If it disagrees with any approach undertaken by EPA, it can override the agency's decision, or respond as it did with acid rain and stratospheric ozone depletion—with new statutory authorities.

# End Notes

[1] U.S. EPA and U.S. Department of Transportation, "Proposed Rulemaking To Establish Light-Duty Vehicle Greenhouse Gas Emission Standards and Corporate Average Fuel Economy Standards," 74 *Federal Register* 49454, September 28, 2009.

[2] http://www.epa.gov/climatechange/endangerment.html.

[3] U.S. EPA, "Prevention of Significant Deterioration and Title V Greenhouse Gas Tailoring Rule," signed September 30, 2009, p. 15. Pre-publication copy available at http://www.epa.gov/nsr/documents/GHGTailoringProposal.pdf.

[4] Ibid.

[5] Memorandum from Jonathan Z. Cannon, EPA General Counsel, to Carol M. Browner, EPA Administrator, EPA's Authority to Regulate Pollutants Emitted by Electric Power Generation Sources (April 10, 1998).

[6] The lead petitioner was the International Center for Technology Assessment (ICTA). The petition may be found on their website at http://www.icta.org/doc/ghgpet2.pdf.

[7] The agency argued that it lacked statutory authority to regulate greenhouse gases: Congress "was well aware of the global climate change issue" when it last comprehensively amended the Clean Air Act in 1990, according to the agency, but "it declined to adopt a proposed amendment establishing binding emissions limitations." Massachusetts v. EPA, 549 U.S. 497 (2007).

# Climate Change                                           109

[8] Memorandum from Robert E. Fabricant, EPA General Counsel, to Marianne L. Horinko, EPA Acting Administrator, EPA's Authority to Impose Mandatory Controls to Address Global Climate Change Under the Clean Air Act (August 28, 2003).

[9] Massachusetts v. EPA, 549 U.S. 497 (2007). The majority held: "The Clean Air Act's sweeping definition of 'air pollutant' includes '*any* air pollution agent or combination of such agents, including *any* physical, chemical ... substance or matter which is emitted into or otherwise enters the ambient air....' ... Carbon dioxide, methane, nitrous oxide, and hydrofluorocarbons are without a doubt 'physical [and] chemical ... substances[s] which [are] emitted into ... the ambient air.' The statute is unambiguous."

[10] For further discussion of the Court's decision, see CRS Report RS22665, *The Supreme Court's Climate Change Decision: Massachusetts v. EPA*, by Robert Meltz.

[11] U.S. EPA, "Regulating Greenhouse Gas Emissions Under the Clean Air Act," 73 *Federal Register* 44354, July 30, 2008.

[12] "Regulating Greenhouse Gas Emissions Under the Clean Air Act," 73 *Federal Register* 44356, July 30, 2008.

[13] Ibid.

[14] Environmental Protection Agency, "Proposed Endangerment and Cause or Contribute Findings for Greenhouse Gases Under Section 202(a) of the Clean Air Act," PrePublication Copy, April 17, 2009, at http://epa.gov/climatechange/endangerment/downloads/GHG EndangermentProposal.pdf.

[15] For a legal discussion of these initiatives, see CRS Report RL32764, *Climate Change Litigation: A Survey*, by Robert Meltz.

[16] New York v. EPA,, No 06-1322 (D.C. Cir., September 24, 2007)

[17] The Board rejected the Region's argument that it was limited by an historical agency interpretation to read "subject to regulation" as meaning "subject to a statutory or regulatory provision that requires actual control of emissions of that pollutant." Since EPA has yet to issue a CAA regulation requiring actual control of $CO_2$ emissions, Region 8 argued, BACT for $CO_2$ is not required. Hence, the Board remanded the permit to the Region for it to reconsider whether to impose a $CO_2$ BACT limit. Deseret Power Electric Cooperative, PSD Appeal No. 07-03 (E.A.B. November 13, 2008).

[18] Northern Michigan University Ripley Heating Plant, PSD Appeal No. 08-02 (E.A.B. February 18, 2009).

[19] For more information on Desert Rock's PSD-NSR permit, see http://www.epa.gov/region09/air/permit/desert-rock/.

[20] Complaint at 2, *Environmental Integrity Project v. EPA*, No. 1 :09-cv-0021 8 (D.C. Circuit, filled February 4, 2009).

[21] The argument for 350 ppm is based largely on concern over melting glaciers, polar ice caps, and sea level, not direct public health considerations.

[22] These levels are specified in Table 2 of the petition, at http://www.biologicaldiversity.org/programs/climate_law_institute/global_warming_litigation/clean_air_act/pdfs/Petition_GHG_pollution_cap_12-2-2009.pdf.

[23] 73 *Federal Register* 44367, July 30, 2008.

[24] Ibid., p. 44493.

[25] In addition to using Section 111, in its July 2008 Advance Notice of Proposed Rulemaking EPA discussed at some length the possibility of using Section 129 of the act to regulate GHG emissions from solid waste combustion units. This would seem to be among the more unlikely routes to regulation of GHGs.

Section 129 is structured differently from most of the other CAA authorities discussed here: there is no provision for an endangerment finding, and there is no blanket authority for the Administrator to regulate pollutants that endanger public health or welfare; there is, instead, a specific list of 10 types of pollution for which the Administrator shall establish standards, with no provision for adding pollutants to the list.

110        Larry Parker and James E. McCarthy

Furthermore, waste incineration is a relatively small source of GHG emissions. According to the latest EPA *Inventory of Greenhouse Gas Emissions and Sinks*, incineration of waste emitted 20.8 million metric tonnes of $CO_2$ in 2007, less than 0.3% of total U.S. GHG emissions.

To the extent that Section 129 provides broader authority to the Administrator, it does so by referencing Section 111: "The Administrator shall establish performance standards and other requirements pursuant to Section 111 and this section for each category of solid waste incineration units." Thus, the authority the Administrator has over waste combustion units is addressed in our discussion of EPA's authority over stationary sources in general under Section 111.

[26] The federal focus on new facilities arose from several factors. First, it is generally less expensive to design in to new construction necessary control features than to retrofit those features on existing facilities not designed to incorporate them. Second, uniform standards for new construction ensures that individual states will not be tempted to slacken environmental control requirements to compete for new industry. NSPS was also seen as enhancing the potential for long-term growth, ensuring competitiveness between low and high sulfur coals, and creating incentives for new control technologies. See Senator Edmund Muskie, Senate Consideration of the Report of the conference Committee (August 4, 1977), in U.S. Senate, Committee on Environment and Public Works, *A Legislative History of the Clean Air Act Amendments of 1977* (95th Congress., 2d session; Serial No. 95-15) (1979), vol. 3, p. 353.

[27] Section 115(a)

[28] 73 *Federal Register* 44483, July 30, 2008.

[29] Roger Martella and Matthew Paulson, "Regulation of Greenhouse Gases Under Section 115 of The Clean Air Act," *Daily Environment Report*, March 9, 2009, pp. 12-17.

[30] 73 *Federal Register* 44483, July 30, 2008.

[31] Martella and Paulson, previously cited, pp. 15-16.

[32] Ibid., p. 11.

[33] 73 *Federal Register* 44519, July 30, 2008.

[34] *U.S. Environmental Protection Agency, Technical Support Document for the Advanced Notice of Proposed Rulemaking for Greenhouse Gases; Stationary Sources, Section VII* (June 5, 2008), final draft.

[35] 73 *Federal Register* 44490, July 30, 2008.

[36] Under Sec. 60.44Da(d)(1), the 1997-2005 NSPS is set at 1.6 lb per megawatt-hour gross energy output, based on a 20-day rolling average; it is lowered to 1.0 lb per megawatthour gross energy output for powerplants commencing construction after February 28, 2005 (Sec. 60.44Da(e)(1). Under Section 60.44Da(e)(3), the 2005 NSPS for modified sources is at either 1.4 lb. A fuel-neutral standard is also set for reconstructed powerplants.

[37] Donald Shattuck, et al., *A History of Flue Gas Desulfurization (FGD)—The Early Years*, UE Technical Paper (June 2007), p. 3.

[38] 42 U.S.C. 7411, Clean Air Act, Sec. 11 1(a)(1).

[39] 40 CFR 60.40-46, Subpart D—Standards of Performance for Fossil-Fuel-Fired Steam Generator for Which Construction is Commenced After August 17, 1971.

[40] Margaret R. Taylor, *The Influence of Government Actions on Innovative Activities in the Development of Environmental Technologies to Control Sulfur Dioxide Emissions from Stationary Sources,* Thesis, Carnegie Institute of Technology (January 2001), pp. 37, 40.

[41] Fri v. Sierra Club, 412 US 541 (1973). This decision resulted in EPA issuing "prevention of significant deterioration" regulations in 1974; regulations what were mostly codified in the 1977 Clean Air Amendment (Part C).

[42] Taylor, ibid., p. 37.

[43] Taylor, ibid., p. 39.

Climate Change
111

[44] For a discussion of challenges arising from the early development of FGD, see Donald Shattuck, et al., *A History of Flue Gas Desulfurization (FGD)—The Early Years*, UE Technical Paper (June 2007).

[45] Examples include full-page ads in the Washington Post entitled "Requiem for Scrubbers," "Scrubbers, Described, Examined and Rejected," and "Amen." For an example, see *Washington Post*, p. A32 (October 25, 1974).

[46] 40 CFR 60.40Da-52Da, Subpart Da—Standards of Performance for Electric Utility Steam Generating Units for Which Construction is Commenced After September 18, 1978.

[47] Margaret R. Taylor, Edward S. Rubin, and David A. Hounshell, "Control of $SO_2$ Emissions from Power Plants: A Case of Induced Technological Innovation in the U.S.," *Technological Forecasting & Social Change* (July 2005), p. 697.

[48] Shattuck, et. al., p. 15.

[49] See CRS Report RL33971, *Carbon Dioxide ($CO_2$) Pipelines for Carbon Sequestration: Emerging Policy Issues*, by Paul W. Parfomak and Peter Folger.

[50] See EPA, ANPR, pp. 44514-44516; Lisa Heinzerling, *Testimony Before the Subcommittee on Energy and Air Quality of the Committee on Energy and Commerce*, Hearing (April 10, 2008); Robert R. Nordhaus, "New Wine into Old Bottles: The Feasibility of Greenhouse Gas Regulation Under the Clean Air Act, " *N.Y. U. Environmental Law Journal* (2007), pp. 53-72; Inimai M. Chettiar and Jason A. Schwartz, *The Road Ahead: EPA's Options and Obligations For Regulating Greenhouse Gases* (April 2009); and Alaine Ginocchio, et al., *The Boundaries of Executive Authority: Using Executive Orders to Implement Federal Climate Change Policy* (February 2008).

[51] U.S. Environmental Protection Agency, "Regulating Greenhouse Gas Emissions Under the Clean Air Act; Proposed Rule," 73 Federal Register 44490, July 30, 2008.

[52] Lisa Heinzerling, Testimony Before the Subcommittee on Energy and Air Quality of the Committee on Energy and Commerce, House of Representatives (April 10, 2008), pp. 12-13.

[53] New Jersey v. EPA, 517 F.3d 574 (D.C. Cir. 2008). The case was decided on whether EPA could delist electric generating units as a source of hazardous air pollutants without following the criteria laid out in Section 112(c). For a discussion see CRS Report RS22817, *The D.C. Circuit Rejects EPA's Mercury Rules: New Jersey v. EPA*, by Robert Meltz and James E. McCarthy.

[54] ANPR, p. 44412.

[55] 73 *Federal Register* 44519, July 30, 2008.

[56] The 1977 CAA broadened the air quality control regimen with the addition of the Prevention of Significant Deterioration (PSD) and visibility impairment provisions. The PSD program (Part C of Title I of the CAA) focuses on ambient concentrations of $SO_2$, NOx, and PM in "clean" air areas of the country (i.e., areas where air quality is better than the NAAQS). The provision allows some increase in clean areas' pollution concentrations depending on their classification. In general, historic or recreation areas (e.g., national parks) are classified Class I with very little degradation allowed, while most other areas are classified Class II with moderate degradation allowed. States are allowed to reclassify Class II areas to Class III areas, which would be permitted to degrade up to the NAAQS, but none have ever been reclassified to Class III.

[57] See the Endangerment Finding, footnote 17 (p. 115 of the pre-publication copy), at http://www.epa.gov/ climatechange/endangerment/downloads/FinalFindings.pdf. Also see Prevention of Significant Deterioration and Title V Greenhouse Gas Tailoring Rule, pre-publication copy, September 30, 2009, p. 46, at http://www.epa.gov/nsr/documents/GHGTailoringProposal.pdf.

[58] Ibid. (proposed Tailoring Rule)

[59] Memorandum from EPA Administrator Stephen L. Johnson to Regional Administrators, "EPA's Interpretation of Regulations that Determine Pollutants Covered by Federal

Prevention of Significant Deterioration (PSD) Permit Program," December 18, 2008, 19 pages, at http://www.epa.gov/nsr/documents/psd_interpretive_memo_12.18.08.pdf.

[60] See "Fact Sheet—Reconsideration of Former Administrator Johnson Interpretive Memo on Definition of Pollutants Covered Under the Clean Air Act," at http://www.epa.gov/nsr/fs20090930guidance.html.

[61] Section 169(1) lists 28 categories of sources for which the threshold is to be 100 tons of emissions per year. For all other sources, the threshold is 250 tons. It should be noted that, unlike the definition of major source, the definition of a major modification is defined by regulation, not statute. As defined under the 1970 CAA, a modification is "any physical change in, or change in the method of operation of, a stationary source which increases the amount of any air pollutant emitted by such source or which results in the emission of any air pollutant not previously emitted"(Section 11 1(a)(4)). In subsequent regulations issued in 1975 with respect to NSPS, EPA defined modification as any physical or operational change that resulted in any increase in the maximum hourly emission rate of any controlled air pollutant. EPA regulations also stated that any replacement of existing components that exceeded 50% of the fixed capital costs of building a new facility placed the plant under NSPS, regardless of any change in emissions. With the advent of National Ambient Air Quality Standards non-attainment provisions (Part D), PSD provisions (Part C), and NSR in 1977, a different approach to defining modification was appropriate as the focus was shifted from enforcing NSPS emission rates to achieving attainment and compliance with PSD. In promulgating regulations for the PSD and non- attainment programs, EPA defined "significant" increase in emissions in terms of tons per year emitted by a major source. For sulfur dioxide and nitrogen oxides, the threshold is 40 tons per year. Facilities exceeding that threshold are subject to NSR.

Given this history of setting *de minimis* emission increases for triggering NSR review for modifications, it is possible EPA could set a substantially higher level for at least carbon dioxide emissions, and perhaps other greenhouse gases, if it determined such thresholds were appropriate. In its proposed Tailoring Rule, the agency proposed a threshold of 10,000 − 25,000 tons per year of $CO_2$-equivalent.

[62] Inimai M. Chettiar and Jason A. Schwartz, *The Road Ahead: EPA's Options and Obligations for Regulating Greenhouse Gases*, April 2009, p. 105.

[63] 73 *Federal Register* 44511, July 30, 2008.

[64] Tailoring Rule, previously cited, p. 19.

[65] Cattle Network, November 18, 2008, at *http://www.cattlenetwork.com/Content.asp? ContentID=269579.*

[66] See CRS Report RL32947, *Air Quality Issues and Animal Agriculture: EPA's Air Compliance Agreement.*

[67] 73 *Federal Register* 44512, July 30, 2008. Also see ensuing discussion through page 44514.

[68] Tailoring Rule, previously cited, pp. 1-2.

[69] Inimai M. Chettiar and Jason A. Schwartz, *The Road Ahead: EPA's Options and Obligations for Regulating Greenhouse Gases*, (April 2009), p. 104.

[70] 73 *Federal Register* 44503, July 30, 2008.

[71] 73 *Federal Register* 44507-44511, July 30, 2008; Inimai M. Chettiar and Jason A. Schwartz, *The Road Ahead: EPA's Options and Obligations for Regulating Greenhouse Gases*, (April 2009), pp. 103-106.

[72] Section 304(a)(3).

[73] *The Clean Air Act and Proposed Acid Rain Legislation: Can We Get There from Here?* CRS Report 85-50 ENR, by Larry B Parker, John E. Blodgett, Alvin Kaufman, and Donald Dulchinos, p. 9.

[74] Testimony of Lisa Heinzerling, U.S. Congress, House Committee on Energy and Commerce, Subcommittee on Energy and Air Quality, *Strengths and Weaknesses of Regulating Greenhouse Gas Emissions Under Existing Clean Air Act Authorities*, 110th Cong., 2nd sess., April 10, 2008, pp. 14-15.

In: EPA Regulation of Greenhouse Gases      ISBN: 978-1-61470-729-5
Editors: Cianni Marino and Nico Costa  © 2012 Nova Science Publishers, Inc.

*Chapter 5*

# STATEMENT OF REGINA A. MCCARTHY, ASSISTANT ADMINISTRATOR, OFFICE OF AIR AND RADIATION, U.S. ENVIROMENTAL PROTECTION AGENCY BEFORE THE COMMITTEE ON ENVIRONMENT AND PUBLIC WORKS U.S. SENATE MARCH 4, 2010

## *Regina A. McCarthy*

Chairman Boxer, Subcommittee Chairman Carper, Ranking Member Inhofe, Subcommittee Ranking Member Vitter, and members of the Committee, thank you for inviting me to testify today to update you on EPA's efforts to mitigate the impacts of emissions from power plants. As you will recall, I last appeared before this committee to discuss these issues in July 2009, and since that time I am pleased to report that EPA has made significant progress on our regulatory efforts to address the public health and environmental effects of air pollutants from power plants. In my testimony I will discuss the status of our work on these efforts, and will provide the committee with some information on S. 2995, the Clean Air Act Amendments of 2010.

From the outset of this administration, beginning with the American Recovery and Reinvestment Act, President Obama has made providing clean

114 Statement of Regina A. McCarthy ...

energy for Americans a top priority. Not only is this enterprise essential to protecting public health and the environment, but it also serves as the cornerstone of revitalizing the economy, spurring innovation and creating new 21$^{st}$ century jobs. That is why your leadership on this issue, Senator Carper, and that of the cosponsors of S. 2995 and of this committee is especially important.

As EPA continues the air pollution rulemakings that reflect our commitment to protecting public health and the environment and to heeding our legal obligations and as you, Senator Carper, and your colleagues work to advance your legislation, I believe that our respective efforts can be mutually reinforcing. They not only ensure the pollution reductions needed, but support the President's efforts to clean up our energy supply in a way that is consistent with economic growth.

## NEED TO PROTECT PUBLIC HEALTH AND THE ENVIRONMENT

Every day, the emissions of sulfur dioxide ($SO_2$), oxides of nitrogen ($NO_x$), and mercury from power plants threaten the health and the quality of life for millions of Americans. Power plant emissions account for over half of total U.S. $SO_2$ emissions, about 20% of $NO_x$ emissions, and just under half the airborne mercury emissions.

Emissions of $SO_2$ and $NO_x$ contribute to levels of fine particles (PM2.5) in the atmosphere; $NO_x$ also contributes to the formation of ground-level ozone. The health effects of exposure to elevated levels of fine particles and ozone include premature death, more asthma symptoms in those already suffering from that disease, and respiratory and cardiovascular diseases that are often serious enough to require hospitalization. Emissions of mercury also undergo transformation in the environment, forming methylmercury which builds up in fish, and, in turn, in people and animals who eat mercury-contaminated fish. Methylmercury exposure in the womb can affect children's cognitive thinking, memory, attention, language, and fine motor and visual-spatial skills.

Although current emissions levels of these pollutants continue to pose a danger for public health and the environment, the past 30 years have seen substantial progress in lowering emissions from power plants. In 1980 U.S. power plants emitted 17.3 million tons of $SO_2$. In 1990, the year Congress passed the Clean Air Act Amendments that included the Acid Rain Program, power plants still emitted 15.7 million tons of $SO_2$ and 6.7 million tons of $NO_x$. By 2000 power plant emissions had dropped to 11.2 million tons of $SO_2$ and 5.1 million tons of $NO_x$. By 2009, preliminary data show that power plants emitted just 5.75 million tons of $SO_2$ and 2 million tons of $NO_x$. The Acid Rain Program was – and is – not just protecting our lakes and streams from acid rain, but also protecting millions of Americans and Canadians from the harmful effects of fine particles. One peer-reviewed study found that the benefits of the power plan reductions from acid rain program outweigh the costs by more than 40-to-1.[1]

This kind of progress makes me confident that renewed efforts to bring these pollutants down to the levels needed to protect against premature deaths, childhood asthma attacks, and acid rain can succeed. There is work yet to be done: although all coal-fired power plants in the U.S. now control particulate matter, and many do control mercury, $SO_2$ and/or $NO_x$, many are still operating without advanced controls for $SO_2$, $NO_x$, or air toxics. EPA and the Harvard School of Public Health have estimated that a coal-fired power plant operating without these controls results in premature deaths and illnesses.

As you heard from EPA Administrator Jackson at last week's hearing before this committee on EPA's proposed 2011 budget, we have not yet completed our review of S. 2995. Fortunately, last summer my office conducted an analysis for Senator Carper of several different emission reduction scenarios, some of which were very similar to emission limits in S. 2995. In that analysis, which is available on EPA's website[2], we analyzed emissions, electricity prices, and costs, and estimated likely health benefits. Based on that analysis, and our experience modeling similar emission reduction scenarios, it appears that S. 2995 would likely result in tens of thousands of lives saved and as much as hundreds of billions in monetized benefits each year, especially when compared to a base case without major new regulation. These benefits are significantly greater than the estimated costs of implementing the reductions required by the scenarios.

## CLEAN AIR AND THE ECONOMY

History clearly demonstrates that the economy can grow while we clean up the air. Since 1980, overall pollution emissions have been reduced by 54%. Meanwhile, VMT, energy use, and population growth have grown steeply and U.S. GDP, adjusted for inflation, has increased 126 percent. The benefits of reducing air pollution are not academic; they have a real effect on how we live

116 Statement of Regina A. McCarthy ...

and what we spend our money on. Less air pollution from power plants means we can spend less on health care for things like asthma attacks, or hospitalizations and emergency room visits for cardiac or respiratory illnesses. It can mean more days at work and fewer employee sick days. Reducing air pollution from power plants can mean we will be able to enjoy more sweeping vistas at national parks like Great Smoky Mountains National Park, or to eat freshwater fish from a New England lake with less concern for possible mercury contamination.

A Congressionally-mandated 1999 EPA study, which went through extensive peer review, found that for all Clean Air Act programs combined, the benefits from 1990 to 2010 would outweigh the costs by 4-to-1. According to OMB's 2009 "Thompson Report" summarizing the annual costs and benefits of federal regulations, the benefit/cost ratio for EPA air rules between 1998 and 2008 was better than for any other government programs.

Like you, I know that air pollution is not the only thing affecting American families. Jobs are hard to come by, businesses large and small are struggling to get the credit they need, and for many people the economic future looks dimmer than the past. In fact, some people are concerned that the U.S. cannot afford to make the investments we need to clean up our air, or that now is the wrong time to make these investments, or that making these investments will hurt our ability to compete in the global economy.

President Obama, Administrator Lisa Jackson and I disagree with that thinking. Making investments in our existing energy sources, updating them to create a clean and efficient energy infrastructure, and making investments that create jobs here in America, all while reducing the number of people who get sick and the resulting costs to our economy, are, in fact, essential to competing in the global economy.

## EPA's Plans

As you know, both the Clean Air Act and recent rulings by the District of Columbia Circuit Court of Appeals require EPA to complete a series of rulemakings to reduce air pollution from power plants. My testimony here last summer made it clear that EPA plans to take smart and effective actions to do this.

EPA will soon propose a rule to replace the Clean Air Interstate Rule (CAIR). This rule will reduce interstate transport of $SO_2$ and $NO_x$ emissions in the eastern half of the U.S. to help states meet the current health-based air quality standards for fine particles and ozone. This keeps us on target to meet the two-year schedule we informed the court we would be following to replace CAIR following the D.C. Circuit's remand. Working within the framework of the 2008 court decision, we are developing a new approach to reduce regional interstate transport of these long- distance pollutants while guaranteeing that each downwind non-attainment and maintenance area is getting the reductions it is entitled to under the law. Past analyses show that benefits of reducing $SO_2$ and $NO_x$ emissions from power plants in the eastern United States far exceed the costs. In addition to these benefits, we anticipate that many of the emission control technologies installed will also help sources meet their maximum achievable control technology (MACT) air toxics requirements.

Similarly, following action by the same court on the Clean Air Mercury Rule (CAMR) as well as our legal obligations, EPA is developing a rule establishing §112(d) MACT standards for toxic air emissions from power plants, including mercury and acid gases. As you know, the MACT program requires us to set our standards for existing sources at a stringency level reflecting the reductions achieved by the top performing 12% of sources.

When I testified in front of you last summer, I was joined on the panel by John Stephenson, Director of Natural Resources and the Environment at GAO, who testified about their analysis of mercury control technology in the power sector. That GAO report, now final, states that "commercial deployments and 50 DOE and industry tests of sorbent injection systems have achieved, on average, 90 percent reductions in mercury emissions."[3] We are still gathering the information we need to determine what the level of our MACT standard will be; we believe that some coal-fired power plant boilers have already reduced their mercury emissions by 90%. Some have been able to make even larger reductions.

I have committed to you that I will follow the data EPA is now collecting when setting the utility MACT standard; that, after all, is what the law requires. Once the rule is finalized, the Clean Air Act requires MACT controls be installed on existing sources within three years, with the possibility of a one-year extension for specific sources under some limited circumstances. New sources must meet the standards when they begin operations. EPA intends to propose these standards for both new and existing coal- and oil-fired power plants by March 2011.

Since I testified before this subcommittee last year, we have revised the national ambient air quality standards (NAAQS) for nitrogen oxides, proposed to revise our $SO_2$ NAAQS, and proposed to strengthen the ozone NAAQS. As the law requires, EPA's NAAQS decisions are based on sound science and our obligation to protect public health. We anticipate promulgating a final $SO_2$ NAAQS by June and a final ozone NAAQS by August. The States are required through their state implementation plans or SIPs to meet the new NAAQS, and address interstate transport of pollution that contributes to downwind nonattainment or maintenance areas for these standards. On top of any federal requirements, these SIPs could well require additional emissions reductions from power plants over the next decade.

## CLOSING

I am confident that whether it is through legislation like S. 2995 or the Clean Air Act regulations that EPA is developing, reductions in power plant pollution will drive smart investments in pollution control and energy efficiency, as well as in innovative generation technologies, all of which will pay back the American people in jobs, economic growth, better health, and environmental protection for years to come.

One of my top priorities at EPA is to work with you, with the power industry, with the states, with community groups and environmental groups, and with the full range of experts from government, business, and universities to find the right path forward in crafting the laws and regulations needed to protect human health and the environment. In closing, I would like to thank Senator Carper and other members of the committee for your strong leadership on these issues over the years. I am confident that we can make great strides to meet our shared environmental and economic goals.

Thank you. I look forward to answering your questions.

## End Notes

[1] Chestnut and Mills, 2005, A fresh look at the costs and benefits of the U.S. Acid Rain Program, Journal of Environmental Management, vol. 77(3):252-266

[2] www.epa.gov/airmarkets/progsregs/cair/docs/CABriefing.ppt

[3] GAO, 2009. Mercury Control Technologies at Coal-Fired Power Plants Have Achieved Substantial Emissions Reductions GAO 10-47.

In: EPA Regulation of Greenhouse Gases     ISBN: 978-1-61470-729-5
Editors: Cianni Marino and Nico Costa   © 2012 Nova Science Publishers, Inc.

*Chapter 6*

# STATEMENT OF REGINA A. MCCARTHY, ASSISTANT ADMINISTRATOR, OFFICE OF AIR AND RADIATION, U.S. ENVIROMENTAL PROTECTION AGENCY BEFORE THE SUBCOMMITTEE ON CLEAN AIR AND NUCLEAR SAFETY COMMITTEE ON ENVIRONMENT AND PUBLIC WORKS U.S. SENATE JULY 9, 2009

## *Regina A. McCarthy*

Chairman Carper, Ranking Member Vitter, and members of the Subcommittee, thank you for inviting me to testify today about EPA's efforts to mitigate the impacts of emissions from power plants. During my visits with many of you during my confirmation process to be the head of EPA's Office of Air and Radiation, I appreciated the opportunity to discuss with you our shared concerns over the public health and environmental effects of air pollutants from power plants. I agree with Senator Carper that emissions of $SO_2$, $NO_x$, mercury, and other pollutants from the generation of energy is a cause for great concern, and I am grateful for his leadership on this important

120 Statement of Regina A. Mccarthy ...

issue over the years. I am glad that we have begun this dialogue and I look forward to continuing to work on this issue.

As I stated at my confirmation hearing in April, I take my responsibility to protect our health and our environment very seriously and I care deeply about these issues. I know everyone here is familiar with the range of serious health and environmental problems caused by $SO_2$, $NO_x$, and mercury. For over a generation we have been hoping and expecting that we could lower emissions from power plants enough to dramatically reduce the frequency of problems like premature deaths, childhood asthma, and acid rain. We have made great progress since 1970, but we have a great distance yet to go. In the meantime, we have learned about the likelihood of increasing serious public health and environmental risks due to CO2 and other greenhouse gases, and we must now make complex decisions about how to address that threat.

In 1980 U.S. power plants emitted 17.3 million tons of $SO_2$. In 1990, the year Congress passed the Clean Air Act Amendments that included the Acid Rain Program, power plants still emitted 15.7 million tons of $SO_2$ and 6.7 million tons of $NO_x$. By 2000 power plant emissions had dropped to 11.2 million tons of $SO_2$ and 5.1 million tons of $NO_x$. The Acid Rain Program was – and is - not just protecting our lakes and streams from acid rain, but also protecting millions of Americans and Canadians from the harmful effects of fine particles. In 2008, power plants emitted 7.6 million tons of $SO_2$ and 3 million tons of $NO_x$. While all coal-fired power plants in the U.S. now control particulate matter and about two-thirds of them use advanced pollution controls for $SO_2$ and/or $NO_x$, it is clear that more cost-effective emission reductions are both necessary and possible.

Data from the 2005 Clean Air Interstate Rule (CAIR) show that the power industry is capable of making significant reductions quickly. The data show that in just three years since the rule was finalized in 2005, the sources planning to comply with it reduced annual $SO_2$ emissions by 2.5 million tons and annual $NO_x$ emissions by 500,000 tons, with many of these emission reductions taking place in the critical summertime ozone season. Rest assured that even though CAIR has now been remanded to EPA, we do not intend to backslide and lose the public health or environmental benefits we have worked so hard to get. Instead, we intend to use what we learned from the CAIR experience to move forward.

And so we find ourselves facing both the great responsibility and the great opportunity to reshape our approach to reducing air pollution from power plants. When I arrived at EPA last month, I was greeted with a briefing that outlined about a dozen pending or imminent rules and decisions affecting

## Statement of Regina A. Mccarthy ... 121

power plants. It is a little overwhelming to be sure, but taken together, these pieces will pave the way towards significant progress over the next few years. Our goal is to make these regulations and decisions work together in a coordinated way to reduce emissions from the entire industry. My purpose here today is to support your efforts to reduce emissions from power plants, to confirm that cost-effective reductions are necessary and possible, and to talk about what we are doing under existing law to achieve these common goals.

Before I arrived at EPA this spring, Administrator Lisa Jackson signed a proposal to regulate mercury and other toxic air emissions from cement kilns. While this rule does not affect power plants, I mention it today because I believe it sends the message that this Administration is serious about reducing air pollution. Since my nomination and confirmation, Administrator Jackson has made it clear that she wants us to move forward both smartly and aggressively. Smartly means we take care to make our regulations fit as best we can into the existing regulatory and economic landscape, that they make sense, that they offer what flexibility they can without sacrificing human health or environmental protection, and that they are cost-effective. Aggressively means we protect public health and the environment as much as we can and as soon as we can.

Recently, I took the first step in implementing this approach by issuing a notice of the Agency's intent to collect information about power plants The information collection would require power plants to provide EPA with data on their emissions of toxic air pollutants, including mercury, acid gases, and dioxin. This data collection could be extensive, but it is necessary for us to set smart and aggressive Maximum Achievable Control Technology (MACT) standards for toxic air pollutants from power plants. As we receive the data, we plan to analyze it and propose MACT standards for power plants as quickly as our understanding of the issues allows.

As you know, the MACT program requires us to set our standards for existing sources at least as stringently as the top performing 12% of sources. Until we identify the top performing 12% of sources, we will not know what that level will be. I can tell you that there are some coal-fired power plant boilers that have already reduced their mercury emissions by 90% or more. The engineers at EPA tell me it is likely the data we collect will indicate the need for a MACT standard that reduces emissions of many air toxics by similar amounts. I have committed to them and to you that I will follow where the data lead on this and all issues. I can also tell you that the MACT program requires these controls to be installed on existing sources within three years

122 Statement of Regina A. Mccarthy ...

after the rule is finalized, with the possibility of an extension of another year for specific sources under some limited circumstances.

EPA is also continuing to address the problem of interstate transport of $SO_2$ and NOx emissions and the resulting fine particle and ozone pollution across the eastern U.S. Working within the framework of the 2008 court decision that remanded CAIR, we are developing a new approach to reduce regional interstate transport of these long-distance pollutants while guaranteeing that each downwind non-attainment area is getting the reductions it is entitled to under the law. We have told the court that we thought it would take about two years to develop a final rule to replace CAIR. We are well underway with the necessary emissions and air quality modeling, and both staff and managers have already held many meetings with various stakeholders, particularly the states, so that we can consider their perspectives early in our process. Although this transport rule may be finalized before the MACT standard, it will take into account any relevant emissions data from the Information Collection Request I mentioned earlier. In addition, it is likely that whatever controls end up being necessary to meet our MACT standard are also likely to substantially reduce emissions of $SO_2$ and particulate matter (PM).

There are other rules EPA is working on that affect power plants – rules that reduce regional haze, rules that help states achieve the National Ambient Air Quality Standards, and rules that reduce emissions from new or modified power plants. Each of these rules is an important component of our ability to clean up the air we breathe and improve the health of the world around us, and will complement the MACT and interstate transport rules.

For example, in January of this year EPA made findings of "failure to submit" for 38 states that had not submitted regional haze State Implementation Plans (SIPs). This started the "clock" for us to put a Federal Implementation Plan (FIP) in place by January of 2011. In addition, none of the regional haze SIPs we have received have been approved, in most cases because they rely on reductions from CAIR that may not be enforceable in the future. We are currently working to coordinate our modeling and decision-making on these haze SIP and FIP issues with the decisions we are making for the utility MACT and interstate transport rules.

We are also using our non-regulatory tools to help reduce emissions by reducing energy demand and supporting the transition to a clean energy future. EPA and the Department of Energy are working together on Energy Star, which helps individuals, companies, cities, counties, states, and the federal government to reduce their energy use. Energy Star has helped revolutionize the market for cost-effective energy efficient products, and provides a wide

## Statement of Regina A. Mccarthy ... 123

range of tools and resources to help homeowners and businesses reduce their energy costs. For example, "Home Performance with Energy Star" has 27 sponsors who have completed over 50,000 retrofits to date that can serve as a model for reducing greenhouse gas emissions from the residential buildings sector. EPA also has programs that provide information to state and local governments about different ways to design successful energy efficiency and renewable energy programs.

The Administration is a strong supporter of energy efficiency across the board. EPA is working with federal agencies such as DOE and HUD that received American Recovery and Reinvestment Act (ARRA) funding for weatherization to install the best types of energy efficiency upgrades, as well as to address other environmental concerns that can come up during these renovations such as indoor air quality and lead contamination. We also continue to coordinate with other federal agencies that have responsibilities for power generation, such as DOE and FERC, to help make energy efficiency a top priority energy resource.

State and federal agencies are working to reduce demand for electricity in dozens of other complementary ways as well, such as supporting policies to realign utility business models so that investing in energy efficiency is no longer a disincentive, supporting demand-response programs and related pricing structures, weatherizing homes, supporting combined heat and power, and installing the beginnings of smart grid technologies. DOE is investing in clean energy technology development, and working with cities and counties to experiment with innovative financing to help customers install them. All of these efforts and many more are helping us clean the air by making it possible to imagine a not-too-distant future where we can shift investments and subsidies away from the highest polluting plants that are the hardest to clean up and towards the clean state-of-the-art technologies we need for the future.

Let me share a quick example from my experience in Connecticut that illustrates the importance of including demand-side tools in our efforts to reduce air pollution from power plants. Connecticut is making an aggressive commitment to using energy efficiency as a priority energy resource by employing a full suite of strategies. These strategies include requiring that a growing percentage of energy demand be met through energy efficiency, identifying energy efficiency as the resource of first choice in energy planning, aggressive funding of state and utility run programs, and rewarding utilities for their achieved energy savings. Among other things, this focused effort on energy efficiency has allowed the state to target investment in Southwest

124 Statement of Regina A. Mccarthy …

Connecticut, a load pocket that includes part of New York City, which was under intense scrutiny to ensure adequate reliability.

The six New England states, the Independent Transmission System Operator (ISO), and industry stakeholders have worked together to develop something called a "Forward Capacity Market." Under this system, ISO New England can project the needs of the power industry three years in advance, and then hold an auction to purchase the resources – either demand- or supply-side resources - necessary to meet those needs. In the latest auction, in December of 2008, over 2,900 MW of demand-side resources "cleared;" that is 400 MW more than cleared in the first auction. Briefly put, this means it is cheaper to save this electricity than to build new capacity to generate it, and that states and industry are continuing to improve their ability to draw on energy efficiency as a valuable resource. These demand-side resources are pollution-free megawatts that make it easier for our states to meet their clean air obligations and – most importantly – for our citizens to breathe clean air.

No discussion regarding the future of the power industry is complete without discussing greenhouse gas pollution control. As Administrator Jackson has repeatedly said, the best approach would be to address this through comprehensive energy legislation. Like many of you, I watched the debate in the House over the Waxman-Markey bill closely, and EPA staff provided timely modeling to assist the legislators there. In addition, we are laying the groundwork for new climate legislation and regulation through efforts such as the Mandatory Reporting Rule. There are of course still some unanswered questions about what exactly the final bill will require power plants to do, but I am confident of one thing: the new crop of power plants will look very different than they do now. I have no illusions that this transition will be easy, but I know it is necessary. I am confident that the laws we adopt and the regulations we implement will drive smart investments in pollution control and energy efficiency, as well as innovative generation technologies, that will pay back benefits for years to come.

As I consider how best to move forward to protect public health, meet our legal obligations, and support this transformation of the energy industry, I come back to the following basic principles. First, we must get emission reductions at the appropriate global, regional, and/or local scales as quickly as is practical. This means getting our new policies in place as quickly as we can while doing what we can to keep emissions from rising in the meantime. These policies must cover all the pollutants we are responsible for: $SO_2$, $NO_x$, air toxics (e.g., mercury, acid gases, others), and greenhouse gases. Second, we must not pay any more than necessary to reach our environmental goals. This

## Statement of Regina A. Mccarthy ... 125

means looking for cost-effective ways to get reductions on the right geographic scales by using the right combination of emissions trading, performance standards and hybrid approaches as appropriate; providing industry the kind of information they can rely on to plan for the future so we can keep the lights on *and* make smart investments; and avoid unnecessarily high or volatile energy costs for consumers. Third, our policies must be clear, coordinated, and legally defensible. Finally, we must keep in mind that soon we will likely be living in a carbon controlled world that will require greenhouse gas emission reductions from power plants. As we plan for the future, it is both environmentally and economically irresponsible not to take this likelihood into account.

To sum up, we are working hard to coordinate our approach to regulating power plants both now and in the future. I am not saying we will solve all of our problems next year, or the year after, but I am saying we are committed to this effort just as you are. One of my top priorities at EPA is to work with you, with the power industry, with community groups and environmental groups, and with experts from government, business, and universities to find the right path forward. This path will let us meet our legal and moral obligations to protect human health and the environment and keep the lights on, all while laying a strong foundation for future changes and investments in the years to come.

In closing, I would like to thank Senator Carper and other members of the committee for your strong leadership on these issues over the years. I am confident we can make great strides to meet our shared goals to protect public health and the environment from the effects of air pollution in the near future.

Thank you. I look forward to answering your questions.

# INDEX

## #

21st century, 113

## A

access, 40
accommodations, 61
accounting, 74
acid, 5, 19, 37, 79, 87, 106, 107, 108, 114, 115, 117, 119, 120, 121, 124
ADA, 61
adjustment, 30
administrators, 107
Advance Notice of Proposed Rulemaking (ANPR), 31, 52, 64, 77
Advanced Energy Projects, 15
advancements, 30
adverse effects, 81, 87, 92
Africa, 90
agencies, 6, 10, 26, 34, 60, 100, 104, 122, 123
agency decisions, 59
air emissions, 117, 120
air pollutants, viii, ix, 2, 4, 9, 19, 26, 33, 46, 50, 51, 52, 74, 77, 80, 81, 85, 86, 87, 100, 102, 103, 105, 111, 113, 119, 121
air quality, 19, 95, 106, 111, 116, 121, 123
air quality model, 121
air toxics, 86, 115, 117, 121, 124

Alaska, 62, 71
ambient air, 9, 17, 70, 81, 82, 87, 89, 107, 109, 117
American Recovery and Reinvestment Act, 113, 122
Americans with Disabilities Act, 61
analytical framework, 35, 39
appropriations, 2, 4, 10, 16, 20
Appropriations Act, 2, 16
assessment, 32, 34
asthma, 114, 115, 119
asthma attacks, 115
atmosphere, 60, 62, 82, 84, 114
attachment, 71
authorities, viii, 4, 16, 17, 18, 24, 27, 33, 34, 35, 37, 38, 39, 41, 42, 45, 50, 56, 67, 73, 74, 75, 76, 77, 78, 97, 99, 101, 104, 108, 109
authority, viii, 3, 8, 9, 10, 11, 14, 16, 17, 18, 20, 24, 36, 39, 40, 45, 49, 50, 51, 55, 56, 57, 60, 62, 65, 66, 73, 74, 77, 81, 86, 87, 88, 92, 98, 99, 100, 103, 104, 105, 108, 109, 110
Automobile, 20, 70
automobiles, 3, 63

## B

background information, 46
banking, 99

barriers, 33
base, 115
Baucus and Stabenow-Brown amendments, 15
beef, 103
benchmarks, 40
benefits, 41, 115, 116, 117, 118, 120, 124
Best Available Control Technology (BACT), viii, 1, 3, 6, 23, 25, 26, 79, 100
bioaccumulation, 87
biofuel, 72
biomass, 30, 31, 34, 43, 93
boilers, 37, 105, 117, 121
business model, 123
businesses, 74, 116, 122

## C

Cabinet, 31, 70, 78
candidates, 35
carbon, 13, 19, 20, 24, 25, 27, 29, 31, 32, 34, 38, 43, 44, 45, 49, 50, 51, 53, 60, 64, 66, 67, 69, 73, 74, 75, 79, 81, 84, 93, 95, 97, 112, 124
carbon dioxide, 13, 19, 20, 25, 27, 29, 32, 34, 44, 45, 49, 51, 60, 73, 75, 81, 84, 93, 112
carbon emissions, 64
carbon monoxide, 20, 60, 81
carbon neutral, 34
cardiovascular disease, 114
case law, 33
catalyst, 44
category a, 105
category b, 29
cattle, 104
challenges, 76, 97, 111
chemical, 70, 86, 91, 94, 109
chemicals, 5, 53, 76, 86, 91
childhood, 115, 119
children, 114
China, 60
chlorine, 43
cities, 122, 123
citizens, 124

City, 44, 48, 55, 123
civil action, 106
classes, 51, 56, 61, 62, 65, 77, 87
classification, 19, 111
clean air, 81, 124
Clean Air Act, v, vii, viii, 1, 2, 3, 4, 6, 8, 10, 13, 14, 15, 16, 17, 18, 19, 20, 23, 25, 26, 31, 33, 35, 36, 46, 47, 48, 49, 50, 51, 52, 55, 56, 60, 64, 65, 66, 68, 69, 70, 71, 73, 74, 75, 76, 77, 78, 81, 90, 91, 95, 97, 100, 103, 105, 106, 107, 108, 109, 110, 111, 112, 113, 114, 116, 117, 118, 120
clean energy, 51, 79, 113, 122, 123
climate, viii, 2, 13, 14, 15, 16, 17, 18, 41, 49, 50, 51, 60, 62, 64, 70, 71, 74, 77, 79, 81, 82, 83, 88, 90, 91, 108, 109, 124
climate change, viii, 2, 13, 14, 15, 41, 49, 51, 60, 64, 77, 88, 90, 91, 108
Clinton Administration, 51
CO2, 5, 7, 13, 19, 23, 25, 31, 32, 40, 42, 45, 48, 49, 51, 53, 54, 57, 59, 60, 61, 62, 63, 68, 71, 76, 77, 78, 79, 81, 82, 83, 86, 93, 94, 97, 100, 102, 104, 109, 110, 111, 112, 120
coal, 2, 8, 23, 24, 25, 29, 30, 31, 32, 38, 39, 40, 42, 43, 44, 45, 79, 93, 94, 95, 96, 104, 115, 117, 120, 121
coatings, 28, 30
cogeneration, 79
collateral, 41
combustion, 5, 28, 29, 31, 45, 75, 76, 93, 109, 110
commercial, 6, 19, 34, 37, 40, 71, 95, 103, 117
community, 34, 118, 125
competitiveness, 110
compilation, 46
complement, 122
compliance, 35, 36, 54, 57, 60, 65, 66, 83, 85, 99, 112
composition, 61
conditioning, 53
conference, 110
configuration, 32, 94

Congress, vii, viii, 2, 3, 4, 5, 6, 10, 11, 12, 13, 15, 16, 17, 18, 20, 49, 50, 51, 69, 70, 73, 74, 75, 79, 86, 90, 92, 96, 98, 101, 102, 107, 108, 110, 112, 114, 120
congressional hearings, 15
consensus, 31, 33, 36, 60, 70, 78
consent, 12
construction, 26, 31, 32, 46, 54, 62, 79, 93, 94, 97, 110
consumers, 124
consumption, 60, 61, 66, 71
contamination, 116, 123
contingency, 98
control measures, 24, 30, 37, 41, 45, 61, 84, 99
controversial, 11, 15
cooling, 45, 62
corporate average fuel economy, 52
correlation, 82
cost, 5, 24, 34, 35, 37, 40, 41, 42, 45, 57, 64, 65, 76, 85, 88, 95, 97, 99, 102, 103, 107, 116, 120, 121, 122, 124
cost effectiveness, 41
cost saving, 97
Court of Appeals, 79, 116
covering, 105
crop, 124
crops, 81, 82
customers, 123

## D

damages, 107
danger, 114
data collection, 121
database, 34
deaths, 115
degradation, 19, 111
denial, 51, 77
Department of Energy, 70, 85, 122
Department of Transportation, 11, 69, 108
deployments, 117
deposition, 87
diesel engines, 58
diesel fuel, 62, 66

dioxin, 121
District of Columbia, 55, 116
draft, 31, 36, 44, 46, 47, 70, 78, 110
durability, 54

## E

economic growth, 114, 118
economic incentives, 99
economic landscape, 121
economic values, 81, 83
economics, 42, 97
effluent, 45
electricity, 40, 44, 65, 72, 115, 123
emergency, 71, 115
emission, vii, 1, 2, 8, 10, 14, 15, 17, 19, 23, 25, 26, 27, 28, 30, 32, 36, 42, 44, 49, 50, 51, 52, 54, 56, 57, 60, 61, 63, 64, 65, 66, 67, 68, 69, 74, 75, 77, 84, 85, 86, 87, 88, 91, 94, 95, 98, 99, 100, 112, 115, 117, 120, 124
emitters, 2, 4, 56, 73
endangered species, 34
energy, ix, 6, 9, 24, 26, 28, 30, 32, 34, 35, 36, 38, 39, 40, 41, 42, 43, 46, 58, 61, 62, 64, 66, 88, 94, 95, 100, 110, 114, 115, 116, 118, 119, 122, 123, 124
energy efficiency, 24, 30, 34, 35, 38, 39, 40, 42, 43, 118, 122, 123, 124
Energy Independence and Security Act, 56, 67
energy supply, 114
enforcement, 8, 83, 108
engineering, 102
England, 123
environment, 43, 91, 113, 114, 118, 119, 121, 125
Environmental Appeals Board (EAB), 38, 79
environmental control, 110
environmental effects, viii, ix, 37, 41, 86, 87, 113, 119
environmental impact, 9, 34, 35, 36, 41, 88, 95
environmental organizations, 64

# Index

environmental protection, 118, 121
epidemiology, 82
equipment, 28, 36, 54, 62, 63, 64, 94, 97, 98
EU, 107
Europe, 59
European Union, 65, 90, 107
evidence, 11, 40, 106, 107
exclusion, 7, 28
Executive Order, 21, 111
exercise, viii, 49, 51, 74
exposure, 82, 86, 114
extraordinary conditions, 57

## F

fairness, 108
families, 116
farms, 74
federal agency, 10
federal government, 122
Federal Register, viii, 12, 18, 19, 20, 31, 46,
52, 70, 71, 72, 73, 74, 77, 78, 108, 109,
110, 111, 112
federal regulations, 116
fish, 114, 116
fishing, 60
flexibility, 9, 37, 87, 88, 90, 99, 100, 101,
103, 121
flue gas, 94, 95
fluidized bed, 43
food, 5, 76
force, 6, 10, 57, 79, 81, 94, 96, 106
Ford, 11
formation, 65, 114
fouling, 62
freshwater, 116
fuel cell, 66
fuel consumption, 58, 59, 60, 61
fuel efficiency, 56, 59
funding, 16, 122, 123
funds, 15

## G

GAO, 117, 118
gasification, 30, 31, 38, 79, 93
GDP, 115
GHG emission standards, vii, 1, 14, 25, 30,
49, 50, 52, 54, 56, 64, 67, 69, 74
global climate change, 70, 108
global economy, 116
global warming, 45, 64, 86
government policy, 96
grants, 15
greenhouse, vii, viii, 1, 2, 3, 4, 5, 6, 8, 9, 13,
14, 15, 16, 17, 19, 23, 24, 25, 26, 27, 29,
31, 33, 37, 44, 45, 46, 48, 49, 50, 51, 52,
53, 56, 60, 64, 66, 67, 69, 70, 72, 73, 75,
76, 77, 78, 81, 82, 84, 85, 86, 88, 89, 90,
91, 92, 93, 97, 101, 103, 105, 106, 107,
108, 112, 120, 122, 124
greenhouse gas (GHG), vii, viii, 1, 23, 49
greenhouse gases, vii, viii, 1, 2, 3, 4, 6, 8, 9,
13, 14, 19, 24, 25, 26, 27, 33, 45, 46, 50,
51, 53, 56, 60, 64, 70, 73, 75, 77, 81, 84,
85, 86, 88, 89, 90, 91, 92, 98, 101, 103,
107, 108, 112, 120, 124
growth, 26, 54, 56, 96, 110
guidance, viii, 4, 23, 24, 25, 26, 27, 28, 29,
30, 32, 33, 34, 35, 36, 37, 38, 39, 40, 42,
43, 44, 45, 47, 75
guidelines, 6, 10, 45

## H

harmful effects, 115, 120
haze, 122
health, 9, 33, 70, 82, 84, 85, 86, 87, 88, 92,
95, 114, 115, 116, 118, 119, 122
health and environmental effects, 85, 87
health care, 115
health effects, 86, 114
history, 86, 112
homeowners, 122
homes, 71, 123
hospitalization, 114

## Index

House, 2, 10, 11, 12, 14, 16, 20, 107, 111, 112, 124
House of Representatives, 111
HUD, 122
human, 82, 86, 87, 91, 118, 121, 125
human health, 82, 86, 87, 91, 118, 121, 125
hybrid, 58, 64, 124
hydrogen, 66

## I

ideal, 88
illusions, 124
improvements, 31, 43, 54, 58, 59, 93, 99
India, 60
individuals, 122
industries, 33, 37
industry, 11, 24, 25, 31, 45, 50, 60, 93, 94, 96, 99, 110, 117, 118, 120, 123, 124, 125
inflation, 59, 115
information sharing, 37
infrastructure, 97, 116
international trade, 65
investment, 30, 123
investments, 116, 118, 123, 124, 125
iron, 37
issues, vii, viii, 2, 3, 6, 18, 25, 26, 31, 33, 42, 59, 65, 70, 78, 81, 113, 118, 119, 121, 122, 125

## J

Japan, 60
jurisdiction, 12
justification, 35

## L

lakes, 114, 120
landfills, 104
laws, 118, 124
laws and regulations, 118
lead, 4, 20, 50, 57, 62, 63, 69, 74, 75, 81, 85, 98, 102, 108, 121, 123

leadership, ix, 113, 118, 119, 125
leaks, 60, 64
legality, 35
legislation, viii, 2, 4, 10, 11, 13, 14, 15, 16, 18, 49, 50, 51, 73, 74, 75, 79, 91, 114, 118, 124
light, 1, 3, 11, 14, 25, 26, 49, 52, 53, 54, 56, 59, 67, 68, 79
light trucks, 1, 3, 25, 49, 52, 53, 54, 67, 69
litigation, 29, 109
local authorities, 37, 38
local conditions, 84
local government, 33, 122
logging, 54
lubricants, 58
lying, 90

## M

machinery, 63
magnitude, 7, 101, 104
major issues, 33
majority, 11, 12, 17, 52, 67, 70, 71, 75, 77, 109
manufacturing, vii, 1, 15, 19, 23, 25, 60, 87
materials, 43, 81, 83
matter, 19, 20, 36, 52, 70, 81, 102, 109, 115, 120, 122
media, 35, 48, 71
median, 54
melting, 109
memory, 114
mercury, ix, 46, 98, 114, 115, 116, 117, 119, 120, 121, 124
Mercury, 111, 117, 118
metals, 5, 76
methodology, 67
Mexico, 55, 79
military, 71
mission, 102
modifications, 19, 61, 102, 112
modified stationary sources, vii, 1, 17, 75
morbidity, 82
mortality, 82

132 Index

motor vehicles, vii, viii, 1, 9, 25, 26, 30, 50, 51, 52, 53, 54, 56, 61, 63, 66, 67, 70, 73, 74, 75, 77, 78, 84, 101, 102

## N

National Aeronautics and Space Administration, 71
National Ambient Air Quality Standards, 16, 17, 46, 81, 112, 122
National Park Service, 71
national parks, 19, 111, 115
National Research Council, 71
natural gas, 24, 30, 32, 39, 43, 44, 45, 94
negotiating, 66
neutral, 32, 46, 62, 93, 110
New England, 116, 123
New Source Performance Standards, 2, 3, 6, 8, 13, 15, 16, 17, 26, 69, 74, 87, 94
nitrogen, 19, 20, 32, 46, 60, 66, 81, 112, 114, 117
nitrogen dioxide, 20, 81
nitrous oxide, 13, 45, 53, 70, 79, 109
North America, 46

## O

Obama, 11, 52
Obama Administration, 11, 52
obstacles, 12
Office of Air and Radiation, v, viii, 72, 113, 119
Office of Management and Budget, 31, 70, 78
oil, 5, 29, 62, 97, 117
omission, 78
operations, 65, 104, 117
opportunities, 26, 29, 40, 60
oxidation, 44
oxygen, 29
ozone, 5, 14, 20, 65, 81, 91, 92, 106, 107, 108, 114, 116, 117, 120, 121

## P

parallel, 20, 74
participants, 17
pathways, 96
peer review, 116
permit, vii, 3, 6, 7, 8, 14, 16, 17, 23, 25, 26, 27, 28, 39, 40, 43, 44, 48, 57, 65, 69, 74, 75, 79, 93, 101, 102, 103, 104, 105, 106, 109
petroleum, vii, 2, 5, 9, 23, 25, 72
Petroleum, 76
phosphate, 19, 87
plants, viii, ix, 19, 24, 46, 69, 79, 87, 94, 95, 96, 113, 114, 115, 119, 120, 121, 122, 123, 124
PM, 19, 111, 122
polar, 109
policy, 4, 5, 6, 31, 70, 75, 76, 78, 108
policy options, 75
policymakers, 107
pollutants, ix, 1, 3, 7, 8, 9, 11, 13, 16, 17, 18, 19, 27, 35, 41, 44, 46, 52, 56, 57, 61, 62, 63, 66, 74, 77, 80, 81, 82, 83, 84, 85, 86, 87, 89, 90, 98, 100, 107, 108, 109, 114, 115, 116, 119, 121, 124
pollution, vii, 1, 2, 3, 4, 8, 19, 23, 25, 26, 30, 40, 51, 52, 54, 55, 56, 57, 60, 61, 62, 65, 70, 71, 77, 83, 84, 85, 87, 89, 90, 94, 105, 108, 109, 111, 114, 115, 116, 117, 118, 120, 121, 123, 124, 125
population, 115
population growth, 115
power generation, 123
power plants, vii, viii, 1, 2, 3, 8, 10, 15, 23, 25, 29, 30, 31, 32, 38, 40, 46, 50, 69, 73, 84, 113, 114, 115, 116, 117, 118, 119, 120, 121, 122, 123, 124, 125
precedent, 61, 105
precedents, 17, 99
precipitation, 107
predicate, 99
premature death, 114, 115, 119
President, vii, viii, 1, 4, 12, 20, 31, 51, 52, 65, 69, 70, 73, 75, 78, 79, 113, 114, 116

# Index 133

President Obama, vii, 1, 51, 69, 79, 113, 116
presidential veto, 11, 13, 20
prevention, 85, 90, 110
Prevention of Significant Deterioration (PSD), vii, 23, 111, 112
principles, 124
project, 26, 30, 34, 39, 41, 44, 123
proposed regulations, 75
protection, 107
prototype, 66
public health, viii, ix, 4, 9, 11, 17, 30, 33, 50, 51, 52, 56, 61, 62, 65, 66, 70, 74, 77, 78, 81, 82, 85, 87, 89, 92, 109, 113, 114, 117, 119, 120, 121, 124, 125
public welfare, 81
pulp, 37

## Q

quality of life, 114
quality standards, 19, 107, 117

## R

radiation, 62, 64
Radiation, v, viii, 19, 72, 113, 119
Ranking Member Inhofe, viii, 113
reading, 2
reasoning, 31, 39, 83, 93, 98, 102
recall, viii, 113
reciprocity, 90
recommendations, 24, 33, 34, 36, 44
recovery, 29, 44, 61, 97
recreation, 19, 111
recreational, 55, 62
reflectivity, 64
regulations, viii, 1, 2, 3, 4, 6, 8, 9, 14, 15, 18, 20, 31, 32, 36, 47, 49, 50, 53, 57, 61, 62, 65, 66, 67, 71, 73, 75, 88, 92, 93, 97, 99, 104, 110, 112, 118, 120, 121, 124
regulatory controls, 11
regulatory requirements, 11
rejection, 28

reliability, 35, 43, 123
relief, 103
renewable energy, 122
renewable fuel, 19, 72, 100
Renewable Fuel Standard, 72
requirements, vii, 2, 3, 4, 5, 6, 7, 8, 9, 14, 15, 16, 17, 18, 23, 25, 26, 27, 29, 32, 33, 35, 36, 41, 42, 57, 62, 63, 65, 69, 74, 75, 76, 79, 84, 88, 89, 90, 91, 94, 95, 96, 98, 99, 100, 102, 103, 104, 105, 106, 108, 110, 117, 118
resistance, 58
resolution, 10, 11, 12, 13, 15, 20
resources, 2, 47, 101, 104, 105, 107, 122, 123
response, viii, 13, 18, 24, 39, 45, 56, 73, 75, 82, 100, 123
RFS, 72
risk, 90
risks, 30, 85, 91, 120
routes, 86, 109
rules, 10, 12, 24, 25, 45, 116, 120, 122
Russia, 60

## S

safety, 61, 62, 64, 65, 81
savings, 123
school, 71
science, 11, 84, 100, 117
scope, 2, 33, 40, 75, 85, 91, 92
sea level, 109
Senate, 2, 10, 11, 12, 14, 16, 17, 20, 110
showing, 57
significance level, 102
smog, 107
snowmobiles, 63
solid waste, 34, 35, 109, 110
solution, 91
speculation, 9, 78, 103
stakeholders, 34, 122, 123
state, viii, 4, 7, 8, 9, 14, 15, 17, 23, 24, 26, 27, 28, 29, 31, 32, 33, 37, 38, 43, 45, 50, 57, 62, 66, 70, 74, 78, 84, 90, 91, 96, 98, 100, 101, 104, 105, 117, 122, 123

# Index

**134**

state authorities, viii, 23, 24, 45
State Implementation Plans, 20, 83, 84, 98, 122
states, 4, 6, 8, 9, 10, 11, 14, 19, 20, 24, 26, 27, 28, 29, 38, 39, 41, 42, 43, 45, 51, 57, 74, 77, 78, 84, 87, 88, 89, 91, 92, 93, 95, 100, 104, 105, 107, 110, 116, 117, 118, 122, 123
statutory authority, 36, 70, 108
statutory provisions, 103
steel, 5, 37
storage, 24, 29, 32, 38, 40, 42, 43, 44, 45, 71, 94, 97
stress, 24, 45
strictures, 90
Subcommittee Ranking Member Vitter, viii, 113
subgroups, 33
substitutes, 91
substitution, 42, 43
sulfate, 62, 107
sulfur, 13, 19, 20, 43, 45, 46, 62, 70, 81, 86, 95, 96, 110, 112, 114
sulfur dioxide, 19, 20, 46, 81, 112, 114
sulfuric acid, 19, 87
Supreme Court, viii, 4, 30, 33, 46, 50, 51, 52, 61, 67, 69, 70, 73, 75, 77, 96, 109
survival, 54
survival rate, 54
symptoms, 114

## T

target, 54, 116, 123
technical support, 32, 93
techniques, 26, 28, 37, 40, 99, 104
technologies, 28, 29, 32, 34, 35, 37, 39, 41, 43, 44, 58, 63, 88, 93, 94, 117, 123
technology, 26, 28, 29, 30, 31, 32, 35, 36, 37, 38, 40, 41, 42, 43, 44, 46, 57, 61, 63, 64, 65, 66, 71, 79, 85, 88, 93, 94, 95, 96, 97, 99, 117, 123
temperature, 43
territorial, 71
thermal energy, 40

time frame, 20
Title I, 19, 50, 51, 56, 68, 80, 99, 107, 111
Title II, 50, 51, 56, 68, 99
Title IV, 107
Title V, viii, 2, 3, 4, 6, 7, 8, 14, 16, 17, 18, 19, 20, 23, 27, 37, 46, 47, 48, 74, 75, 76, 80, 88, 91, 102, 103, 104, 105, 107, 108, 111
trade, 16, 20, 28, 50, 54, 65, 66, 69, 74, 92, 97, 98, 99, 100, 101, 108
training, 34
transformation, 114, 124
transport, 5, 40, 64, 76, 97, 107, 116, 117, 121, 122
transportation, 15, 40, 66, 107
triggers, 5, 74, 89
turnover, 67

## U

uniform, 82, 84, 110
United Nations, 89
United States, vii, 1, 35, 55, 59, 60, 61, 64, 65, 71, 83, 86, 89, 90, 95, 117
universe, 102
universities, 118, 125
updating, 116
urban, 56, 58

## V

variables, 81, 83
variations, 83
vegetation, 26, 34, 81, 82
vehicles, vii, viii, 1, 9, 11, 25, 26, 30, 50, 51, 52, 53, 54, 56, 59, 60, 61, 62, 63, 66, 67, 69, 70, 71, 73, 74, 75, 77, 78, 84, 101, 102
vessels, 55, 60, 61, 71
veto, 11, 13, 15
Vitter, viii, 113, 119
volatile organic compounds, 60
vote, 2, 12, 14, 16, 20

## W

waiver, 14, 35, 36, 37, 43, 47, 57, 69
Washington, 111
waste, 45, 79, 110
waste incineration, 110
water, 26, 34, 35, 45, 65, 81, 82
water vapor, 65
welfare, 4, 9, 11, 17, 30, 33, 50, 51, 52, 56, 61, 62, 65, 66, 70, 74, 77, 78, 81, 82, 87, 89, 92, 109

well-being, 81, 83
White House, 11, 50
wildlife, 81, 82
worry, 108

## Y

yield, 28